58497

Industrial geology

TN
260
.I52

Industrial geology

Industrial geology

EDITED BY
J. L. KNILL

OXFORD UNIVERSITY PRESS

1978

Oxford University Press, Walton Street, Oxford OX2 6DP

OXFORD   LONDON   GLASGOW   NEW YORK
TORONTO   MELBOURNE   WELLINGTON   CAPE TOWN
IBADAN   NAIROBI   DAR ES SALAAM   LUSAKA   ADDIS ABABA
KUALA LUMPUR   SINGAPORE   JAKARTA   HONG KONG   TOKYO
DELHI   BOMBAY   CALCUTTA   MADRAS   KARACHI

© OXFORD UNIVERSITY PRESS 1978

British Library Cataloguing in Publication Data
Industrial geology.
  1. Geology, Economic
  I. Knill, J L
  553   TN260

ISBN 0-19-854520-7

PRINTED IN GREAT BRITAIN
BY RICHARD CLAY (THE CHAUCER PRESS) LTD,
BUNGAY, SUFFOLK

Foreword

THE ACADEMIC training provided by the usual three- or four-year university undergraduate course in geology necessarily concentrates upon the fundamental principles, facts, and concepts of a subject to which a high proportion of students come as beginners. Most university teachers find it difficult enough to deal adequately with the basis of geology within so short a period and many—but by no means all—feel that any attempt to teach much applied geology *per se* in an undergraduate programme is of limited value. Most employers of graduate geologists prefer their recruits to have had a sound training over the broad spectrum of the science and recognize that the professional skill and expertise needing to be gained during the early years of employment cannot be acquired without a really sound background of the principles of pure geology.

At the present time, the academic geologist tends to become carried away by all the latest scientific advances—plate tectonics; the lunar rocks; the geology of the ocean floors—and must take care that all the excitement does not mean that the more ordinary basic arts of a geologist's training are left too thinly covered. Whatever a young geologist is going to do after graduation, he must at university learn his branch of natural history, his fossils, rocks, and minerals. He must acquire as much field experience as humanly possible and learn both to read and to construct a geological map. His subject is evolving so quickly at present that the achievement in a university course of a proper blend of factual knowledge, observation and deduction, and theory presents greater problems now than in the immediate past.

With all this to do, an undergraduate cannot effectively be trained as an applied geologist from the start. This would take five or six years at least, not three, even if it were desirable in principle and the universities had the appropriately qualified staff. But in an increasingly competitive world, students ought to be kept as aware as possible of the potential applications of their discipline in industry, where many will find their subsequent careers. It is not so much a matter of learning applied or industrial geology as such, but more the encouragement of an attitude of mind; the inculcation

v

58497

Foreword

of an awareness that the subject has great practical value, sometimes in unexpected directions.

Until very recently the Department of Geology and Mineralogy at Oxford has had no permanent staff member with the necessary experience to assume the main responsibility for fostering this interest and awareness of practical applications. Feeling this lack acutely, I have followed the policy of organizing, every two or three years, a course of lectures given by various friends and colleagues, largely from industry and Government service who, I am happy to record, have needed remarkably little coercion to come and help us, even though several have insisted bashfully that they were in no sense practised lecturers. These courses have been very successful and our students have profited greatly from our visitors' expositions of 'geology with the lid off', and from being shown convincingly that not only must an industrial geologist know about geology but he must also become something of an economist, something of a manager, something of an engineer—something of so many things in fact that one is easily persuaded to say that breadth of interest and balance and maturity of outlook are more important attributes to be acquired by a young industrial geologist than simple academic brilliance.

The present volume stems from such a series of lectures originally given in this Department in Hilary Term 1974. The initiative for its preparation came from the Oxford University Press, which recognized that there was no single-volume work available for an undergraduate or research student wishing to gain some overall view of the kinds of problem with which an industrial geologist has to deal, and suggested that our lecture course might form the basis for such a book. One of our invited lecturers, Professor John Knill of Imperial College, has undertaken the task of editing the volume, as well as contributing largely to it himself and persuading some contributors not involved in the original lecture course to write special chapters on topics which would not otherwise have been covered. Coal was dealt with in the lecture course by Mr. G. Armstrong, formerly Chief Geologist of the National Coal Board, who did not feel able to constrain his stimulating if unconventional treatment within the more formal mould of a written chapter, and Messrs. Williamson and Barefoot have covered the subject for this book.

Professor G. R. Davis of Imperial College initiated the lecture series with a talk entitled 'The visibility of orebodies' in which his theme was the ways and means by which economic orebodies can be recognized and defined. As the lectures incorporated in this volume include one by Dr. O. W. Nicolls, also on the topic of mineral exploration, Professor Davis has most kindly agreed to produce a new contribution on geology in the minerals industry, in the interests of better balance and broadened scope.

The geological aspects of construction materials were not dealt with in the 1974 lecture course at Oxford and chapters dealing with sand, gravel, and constructional stone, and with cement, have been specially written for this volume by Dr. Diane Knill and Mr. D. P. Jefferson respectively. Similarly, Mr. M. H. de Freitas has contributed a new chapter on geological hazards, and the book is rounded off by a chapter on geology in conservation by Dr. G. P. Black of the Nature Conservancy Council.

Inevitably, some important topics are not covered—industrial minerals, for example—and this book makes no claim to be a comprehensive treatise on industrial geology. Rather, it is a series of essays each written by an expert in the field and differing in approach, emphasis and treatment. It is the hope of editor, authors, and publisher that it will be the kind of volume which a second- or third-year undergraduate in geology will read through and find stimulating and helpful in deciding what kind of career he wants, and to what useful purposes his sometimes rather academically oriented subject can be put. His teachers, too, should read it from cover to cover. The book should also be useful to those who are concerned with the minerals and resources planning fields.

Department of Geology and Mineralogy   E. A. VINCENT
Oxford

Contents

List of contributors   xiii
Abbreviations and units   xv

1. Industrial geology   1
J. L. KNILL

2. The energy resources industries   7
D. C. ION
1. Introduction   7
2. Exploration   7
3. Production   14
4. Reserves   16
5. Planning   21
6. Laws and regulations   24
7. The environment   26
8. Society   30

3. Oil and natural gas   33
H. R. WARMAN
1. Introduction   33
2. The nature and significance of oil in the subsurface fluid regime   33
3. The role of stratigraphy   41
4. The role of structure   45
5. Activities and techniques in the oil industry   47
6. The use of geologists in other roles   49
7. The future for oil and the oil industry   49

4. Coal   54
GEOLOGY APPLIED TO SUBSURFACE MINING   54
I. A. WILLIAMSON
1. Historical   54
2. Geological controls in exploration   56
3. Evaluation of the constraints   60
4. Assessment of reserves   61
5. Exploratory assessment   63
6. The useful geologist   65

GEOLOGY APPLIED IN OPENCAST WORKING   65
M. J. BAREFOOT
7. Historical   65

ix

Contents

8. Investigation 67
9. Extraction 71
10. Reserves 76

5. Geology in the minerals industry 78
G. R. DAVIS

1. The variety of professional demand 78
2. Main lines of activity in mining geology 83
3. Illustrative case-history of Kilembe Mine, Uganda 83
4. Education and opportunity 107

6. Mineral exploration 111
O. W. NICHOLLS

1. Introduction 111
2. Exploration methods 113
3. The exploration programme 123
4. Conclusion 134

7. The metallogeny of Britain 137
K. C. DUNHAM

1. Introduction 137
2. Mineral provinces 138
3. Ages of mineralization 145
4. Structural control of oreshoots 149
5. Temperature control of oreshoots 156
6. Mineralizing fluids and metal sources 160

8. Aggregates, sand, gravel, and constructional stone 166
D. C. KNILL

1. Introduction 166
2. Sand and gravel 168
3. Road aggregates 175
4. Large-scale construction stone 181
5. Aggregates derived from industrial waste and its utilization 184
6. Building stones 189
7. Conclusion 194

9. Geology and the cement industry 196
D. P. JEFFERSON

1. Introduction 196
2. Chemistry of Portland cement 197
3. Manufacture of Portland cement 200
4. Chemical requirements of cement raw materials 203
5. Physical requirements of cement raw materials 205
6. Cement raw materials 206

7. Determination and proving of cement raw materials 213
8. Harmful impurities in cement raw materials 215
9. Blending of raw materials 218
10. Environmental problems 221
11. Conclusion 222

10. Groundwater 222
R. A. DOWNING

1. Introduction 224
2. The nature of aquifers 226
3. Assessment of groundwater resources and development 232
4. Consequences of groundwater development 240
5. Groundwater quality 245
6. Groundwater management 250
7. Conclusion 257

11. Geology in the construction industry 259
J. L. KNILL

1. Introduction 259
2. Investigation 261
3. Design 271
4. Construction 280
5. Operation 284

12. Geological hazards 287
M. H. DE FREITAS

1. Introduction 287
2. A perspective 288
3. Identification of natural hazards 292
4. Identification of induced hazards 294
5. Forecasting 304
6. Future trends 305

13. Geology in conservation 310
G. P. BLACK

1. Conservation 310
2. Geology and conservation 312
3. Conservation and the extractive industries 314
4. Conservation and the construction industries 327
5. Conservation in Britain 330

Further reading 335

Index 339

List of Contributors

M. J. BAREFOOT, Chief Opencast Geologist, Opencast Executive, National Coal Board

G. P. BLACK, Head, Geology & Physiography Section, Nature Conservancy Council

G. R. DAVIS, Professor of Mining Geology, Imperial College, University of London

R. A. DOWNING, Central Water Planning Unit, Department of the Environment

SIR KINGSLEY DUNHAM, *lately* Director, Institute of Geological Sciences. Emeritus Professor of Geology, University of Durham

M. H. DE FREITAS, Lecturer in Engineering Geology, Imperial College, University of London

D. C. ION, *formerly* The British Petroleum Company Ltd., Energy Resources Consultant

D. P. JEFFERSON, Senior Geologist, The Associated Portland Cement Manufacturers Ltd.

D. C. KNILL, St George's School, Ascot

J. L. KNILL, Professor of Engineering Geology, Imperial College, University of London

O. W. NICHOLLS, Senior Exploration Geologist, Selection Trust Ltd.

H. R. WARMAN, *formerly* Exploration Manager, The British Petroleum Company Ltd., Exploration Consultant

IAIN WILLIAMSON, Consultant Geologist, Orrell, nr. Wigan.

Abbreviations and units

THE geologist working on industrial projects will encounter a great variety of units, many of them non-metric, and this situation is likely to persist for many years to come. For this reason it was judged unrealistic to convert all the data quoted in this book to SI units. Metric equivalents, accurate enough for normal purposes, are listed below.

Length

1 inch = 25·4 m
1 ft = 0·3048 m
1 mile = 1·609 km

Area

1 acre = 407 m² = 0·4047 ha
1 square mile = 2·589 km² = 258·9 ha

Volume

1 ft³ = 0·0283 m³
1 yd³ = 0·7646 m³
1 U.S. barrel (42 gallons) = 0·1589 m³ = 158·9 litres
1 U.S. barrel/day = 50 tonnes/year
1 gallon (Imperial) = 4·546 litres
1 million gallons per day = 4·536 Ml/day

Mass

1 ton = 1·016 tonne (metric ton, 1000 kg)
1 lb = 0·4536 kg

Pressure

1 lbf/in² = 6895 pascal (N/m²) = 6·895 kPa (kN/m²)

Rate of flow, permeability

1 cusec (ft³/s) = 0·0283 m³/s
1 darcy = 0·987 × 10⁻⁸ cm²

Abbreviations and units

Energy equivalents

1 U.S. barrel crude oil $= 6.0 \times 10^9$ joules
1 tonne crude oil $= 7.3$ U.S. barrels crude oil $= 44.0 \times 10^9$ joules
1 tonne hard coal $= 28.8 \times 10^9$ joules
1 tonne lignite $= 8.8 \times 10^9$ joules
1 m^3 natural gas $= 3.8 \times 10^7$ joules

Temperature

$1\,°C = 1.8\,°F$

1. Industrial geology

INDUSTRIAL GEOLOGY embraces those aspects of the geological sciences which are applied to the direct benefit of Man. Traditionally professional geology has recognized a boundary, albeit somewhat diffuse, between those who work in education and government service and are engaged in fundamental or strategic research and those working in industry whose responsibilities are concerned primarily with the solution of practical questions coupled with applied research. This distinction between 'pure' and 'applied' geology has never been satisfactory and the feedback from one to the other has made the distinction increasingly unnecessary if not undesirable. One has only to consider the great contribution made by petroleum geologists over the years by the systematic mapping of mountain chains in many parts of the world to appreciate that the actual discovery of oil was a second-order response to the collection and interpretation of geological data. The difference lies not within the discipline of geology but in the attitudes to working within an industry where the restraints can be so very different from those in environments where more academic criteria hold sway. Much of geology is of practical relevance; indeed, it has been stated that all geology is 'applied'. It would, however, be misleading to expect every geologist to carry in his rucksack the skills of locating a borehole for water supply, deciding whether or not a rock face is likely to collapse, and establishing whether surface evidence of mineralization merits costly exploration. The industrial geologist must therefore focus his geological knowledge and then apply it to the question in hand, fully aware of the needs of his client or employer. It is as valid to study an ore deposit as an example of a chemical reaction system within the Earth's crust as it is to study the same ore deposit from the viewpoint of recovering a metal at minimum cost and hazard. Both

1

require considerable geological knowledge if the right answer is to be obtained; the answers are, however, not mutually transferable and there, in part, lies the difference.

The two main industries employing geologists are those associated with extraction and construction. No doubt a debate as to which industry started employing geologists first would rapidly revert to the issue as to whether CroMagnon Man, as he sat in his cave by the Vézère, was more worried by the stability of the limestone roof of his cave or where the next flint was coming from. The vast majority of professional geologists certainly work in the extractive industries and there is little prospect of this proportion altering significantly in the future. An alternative approach is to relate the industrial connection to the ultimate end-product. Thus the energy-based industries would cut across traditional professional groupings, linking together those employed in coal, oil, radioactive minerals, geothermal power, and reservoir construction, among others. Although one can see a progressive convergence of approach, such as the increasing interest in coal resources shown by large oil companies, or the application of petroleum drilling techniques to the exploitation of thermal waters, the geological resource or industrial constraint offers a more logical approach to systematic description. Nevertheless, the wider view is taken by D. C. Ion in Chapter 2 in relation to the energy-based industries.

The extractive industries are those which are associated with the exploitation of oil and natural gas, coals, metallic (ore) minerals, industrial minerals, construction materials, and water. Of these resources only one, groundwater, is renewable at rates determined by both natural and artificial replenishment; groundwater can, however, also be abstracted on a non-renewable basis analogous to the other methods of mineral extraction. The characteristic features of each of these resources may be summarized as follows.

Oil and natural gas. Accumulate in stratigraphic or structural traps in large sedimentary basins which provide suitable source and reservoir rock conditions, and have suffered inadequate erosion to cause loss of hydrocarbons. Development of resource by drilling of production wells. Transitional materials, such as tar sands and oil shales, are developed by bulk mining operations.

Coals. Occur within sedimentary rocks in or marginal to sedimentary basins; the coal series extends from the peats to lignite, brown coals, and

2

bituminous coal to anthracite. Extracted by subsurface and open-pit mining operations.

Metallic minerals. Occur in a wide variety of geological situations forming deposits ranging from highly concentrated accumulations to disseminated bodies with a metallic content little above that of the background values. Extracted by subsurface and open-pit mining operations. Metallic minerals can be obtained from certain brines.

Industrial minerals. Occur in a wide variety of geological situations with characteristics similar to those of the metallic minerals. Commonly, but not exclusively, extracted by bulk mining operations in open pits.

Construction materials. Almost all hard and soft rock materials can yield some form of construction material in appropriate circumstances. Almost exclusively extracted by open-pit methods.

Water. Groundwaters are renewed by precipitation or surface recharge; resources can be developed on a non-renewable basis. Deep groundwaters may be sources of mineralized brines or thermal waters. Extraction is normally from boreholes.

It will be apparent that there is considerable common ground between not only the methods of extraction but also the techniques of resource evaluation.

The construction industry cannot be so simply related to geology. A major field of construction is associated with the development of water resources including, in particular, surface water for generation of power, provision of industrial, domestic, and irrigation supplies, and the control of floods. Urban development and redevelopment, together with intervening transportation routes, to provide services of all kinds, occupy another important facet of the industry. Significant interaction can take place between the extractive and construction industries, particularly when development extends into old mining fields, or mineral extraction takes place in areas where there are existing structures. Certain extraction operations, such as offshore oil or gas fields, may necessitate large-scale construction at a pace and rate of expenditure unfamiliar to those working in the mainstream of construction operations.

The prediction of geological conditions, based upon exploration, is central to the work of the industrial geologist, irrespective of the industry within which he is working. The technology of exploration pp. 7, 47, 63, 67, 84, 111, 213, 261) is common to each field as is the background of uncertainty against which major decisions have to be made and the implications costed. In interpreting the geological conditions account must be taken of the adequacy and validity of the data available, the economic restraints within which the particular development is to proceed, potential environmental constraints, and the engineering procedures necessary to bring the project to satisfactory performance. Inevitably, the industrial geologist does not take sole responsibility for all these activities but is a member of an interdisciplinary team (pp. 7, 98, 259) each member of which contributes a particular expertise or performs a specific function. Such teams will contribute different approaches and experience of alternative disciplines including other sciences, engineering, economics, and law. The geologist must be scientifically versatile, able to communicate effectively, and be motivated towards, and sympathetic to, the objectives of industry. In sum, the geologist finds himself as one member of the team who is especially able to appreciate the interaction between the development, the environment, and society; this may well raise moral questions which can never be easily resolved, and may give rise to personal crises of allegiance and responsibility. It is such questions which distinguish the industrial geologist from his fellow-professionals. Few students have the opportunity of appreciating the fundamental differences between the industrial and academic viewpoints. In industry it is usual to work within a team, to have to achieve tasks within predetermined timetables, to take cost into account in most decisions, and to present results to multidisciplinary or lay audiences requiring simplicity and clarity of expression; such requirements are almost in total antipathy to the traditional concept of the academic environment. There is, in consequence, a distinction to be drawn between 'applied' geology, which can be taught almost divorced from the real world, and industrial geology, which recognizes to the full all the interactions necessary to achieve the end result.

Apart from this common ground associated with the working environment of industrial geologists, there is considerable, and increasing, overlap between the activities of geologists employed within different industries. For example, the development of low-grade ores in disseminated deposits and the increasing interest in large-scale surface extraction of coal has emphasized the need for information on geological conditions which may influence slope stability and groundwater flow. Traditionally the mining or mineral exploration geologist has been concerned primarily with the location and delimitation of resources, but as indicated by G. R. Davis (pp. 98–101) professional needs can require a much wider view of geology within the minerals industry. The development of the hydrocarbon resources of the North Sea has led to the need to develop new well-head structures for production wells and construct pipelines in conditions which have not been evaluated previously for construction purposes. Not only has it been necessary to develop new technology for investigation and construction but the economic restraints of the extractive industry have become imposed upon those of the construction industry. These restraints will mean, in effect, that a more approximate solution, based upon minimal data, provided quickly may be economically more valuable than a solution which is more accurate but takes longer to achieve. This must inevitably throw back questions about the quantity of geological data, and the envelope of uncertainty associated with those data, needed to solve different problems in different industries. Less prosaically, one can draw upon the increasing recognition of the importance of fluid behaviour in the Earth's crust in relation to such diverse subjects as the accumulation of ore minerals (p. 161), the occurrence of brines (p. 247), high formation pressures at depths in excess of about 2000 m (p. 40), the hazards of deep injection of fluids (pp. 250, 301) and possible control of earthquake activity, and the development of geothermal power (p. 13). The recognition of possible associations between lead–zinc sulphide ores and the genesis of hydrocarbons illustrate clearly the multidisciplinary trends within industrial geology. However, these trends must draw upon fundamental or strategic research that is not necessarily directly related to applied, or industrial, geology. For example, the application of plate tectonics theory and the testing of that theory against field observations leads to a greater appreciation of the geochemical evolution of the earth and, in consequence, to a greater likelihood of ores of particular types being found in rocks of particular ages. By such means the preliminary stages of a search for mineral deposits can be focused more rapidly on potential regions. Progressive refinement of our understanding of mountain-building processes will enable progressively smaller areas to be identified during the search process; this is well illustrated by the tectonic controls on the localization of many porphyry copper deposits. Thus, K. C. Dunham's chapter underlines the importance of the recognition of the metallogenic history, not only of individual regions, but the Earth as a whole.

Extraction and construction do not take place in a vacuum, and it is important to appreciate the interactions which take place between these activities, Man, and the environment. Although many geological hazards are the result of natural processes caused by catastrophic events, an increasing number of hazards are related to engineering operations of different types. Equally, the after-use of an abandoned mineral site and the scientific value of new excavations as a geological resource must be appreciated and taken into account during planning and subsequent development.

Industrial geology is an amalgam of geological knowledge, techniques developed to serve the needs of a particular industry, and the ability to work within the constraints required by that industry. This book provides a number of examples of the application of industrial geology in practice and the role of geologists within industry.

2. The energy resource industries

1. Introduction

The part the geologist must play in the future of all the world energy resource fields will probably be greater than it has been in the past. The growing realization that world energy demand is stretching the known non-renewable resources has led not only to greater exploration effort to find new sources but has also emphasized the interdependence of all energy sources to provide the optimum mix to satisfy total demand. For many years the world will need both additions to its known resources and wiser deployment of those resources, using to the maximum the characteristics peculiar to each and yet causing minimum harm to the environment. The scope of applied geology has widened rapidly in the past twenty years and it will continue to widen and deepen in the future.

2. Exploration

The traditional role of the geologist is in exploration, whether for coal, oil, natural gas, uranium ores, geothermal energy sources, or for hydro-electric installation sites.

During the past fifty years, and particularly since the mid-1950s, the style of exploration has changed as increasingly elaborate tools have become available, many of them requiring specialist operators. The lone prospector, or the former two-man survey party, has become outmoded; a much larger team is now required both in the field and in associated back-up services. Reconnaissance mapping of the topography and geology of virgin land with plane table and range-finder has been replaced by aerial photography, remote sensing, interpretation, and computer mapping before the field geologist even goes to selected areas. The helicopter has replaced the horse. Collaboration with geophysicists has increased as skills have improved to measure the physical characteristics of buried rocks, even below the sea-bed, at greater depths and with greater resolutions. Drilling to greater depths, on land and sea, has given more facts and has advanced other technologies. Such techniques as the dating of rocks by their radioactive characteristics, data storage and retrieval systems, mathematical modelling, and greater precision in analytical methods are all new tools in the kit of the explorer. A recent addition is the data received from the Earth Resources Technology Satellite (ERTS). Such imagery, giving a synoptic view of large areas under uniform lighting, can indicate trends or structures which are not recognizable by photography from aircraft. Targets for further investigation may be indicated, e.g. unmapped buried salt domes, which may be petroliferous, as in southern Mississippi, or new exploration areas, as in the Anadarko Basin of western Oklahoma, U.S.A. It is estimated that from 20 to 50 per cent of the cost of a conventional field survey might be saved by prior planning after review of the data from satellite scanning. Canada is developing a 'quick-look capability' to provide data within 24 hours of reception. This has already been of direct help to marine seismic exploration crews in the High Arctic in that it has allowed them to cover greater distances by indicating open water ahead. Another technique being developed is to relay from transmitters monitoring local conditions in remote areas automatically to the base via ERTS. This has already been used effectively in obtaining water resource data for hydro-electric schemes. There will be many more applications of various types of satellite sensing which will be further developed to aid the explorer. However, the most important tools by far are still the geological concepts, hypotheses, and ideas for which all the other tools provide only facts.

It has been argued that progress in exploration for energy resources has been taken in steps, as developments occurred in geological thinking or technical capability, with the greatest advances occuring when these coincided. The petroleum field can offer examples. In the 1920s it was the development of geophysics, coupled with new thinking about salt dome mechanisms, which opened up the Gulf Coast of the United States. In 1932, the discovery of light oil in Bahrain reversed the geological thinking of the 1920s that only heavy oil, if any, would be found west of the Persian Gulf. Up to about 1947 the petroleum prospects shown on world maps were limited to land areas, yet by 1954 offshore areas were included. The technical drilling ability had been developed, in this case gradually, in the swamps of the Mississippi delta and the shallow waters of Lake Maracaibo. The first well drilled out of sight of land was in 1948 off Louisiana. Once the geologist was allowed to look seawards, his frontiers were widened enormously. These new prospects, and the remarkably rapid technical progress made in overcoming drilling and production problems at sea, have been matched by progress in geological thinking about the oceans. Scientific curiosity about the oceans was, indeed, encouraged by the petroleum industry. Out of the Mohole project grew the Joint Oceanographic Institutions Deep Sea Sampling Project (JOIDES), which has added so much to our knowledge of the ocean floor and the build-up of plate tectonic theories.

The fossil fuels and the nuclear fuels are widely scattered through the geological column and are of different ages and lithological types. Although most of the sedimentary basins on land, and now some offshore, have been partially explored, there may well be many surprises in store. Few indeed in 1964 forecast the richness of the oil basin in the northern North Sea, for it was gas that was looked for at that time, not oil. Before 1953 it was accepted that the uranium ores in the rich Colorado–Wyoming area of the U.S.A. were syngenetic, being essentially confined to certain beds of Jurassic age. Dating of the ores, however, showed that some were younger than the host rocks and this led to important additions to the reserves as the search was widened to rocks of all ages in the area.

Coal-mines have been extended from land for some distance beneath the sea-bed but, as yet, only petroleum is commercially exploited far off shore. Perhaps, in the distant future, improved knowledge of plate tectonics may eventually lead to offshore exploitation of submarine hyperthermal areas as heat sources. Because of the masking effect of water, geophysics provides the principal reconnaissance tool off shore, but the interpretation of the measurements made and their combination into a coherent whole is a geological problem. The nature of the sea-bed itself is a geological concern when drilling locations are being decided. Here the oceanographer and the sedimentologist must join the team to forecast the resulting disturbances to the sea bottom which may be caused by the drilling structures.

Coals vary in rank and therefore in commercial attractiveness.

9

The energy resource industries

Crude oils vary even more widely in quality. The coals of lowest rank, the lignites, and the crude oils of lowest quality, the viscous hydrocarbons which cannot be recovered in their natural state through wells by normal production methods, are rarely exploration targets; however, technical innovations may well raise their value. More important, it should be appreciated that, in the past, quantitative evaluation of such deposits has been very sketchy. One published estimate of the amount of heavy oil in place in the world is about 1600×10^9 barrels U.S. (254×10^9 m^3). Another estimate gives 700×10^9 barrels (111×10^9 m^3) to the Orinoco Tarbelt, Venezuela alone. There are many such occurrences of heavy oils in Canada, Venezuela, the U.S.A., Malagasy, Albania, Trinidad, Colombia, Romania, the U.S.S.R., the Philippines, Thailand, Kuwait, and parts of West Africa. Only one plant in the Athabasca Oil Sands of Canada is at present exploiting these oils to any significant extent.

Richness and size are of paramount importance to all energy resource deposits. Some 125 very large oilfields contain 70 per cent of the world's proved recoverable reserves of crude oil. Roughly 50 per cent of the world's known remaining recoverable natural gas is in 20 fields, with reserves varying from 12 to 20×10^{12} ft^3 (34 to 57×10^{10} m^3). Three uranium ore deposits contain roughly 50 per cent of the proved reserves, recoverable at a price less than $20/kg U_3O_8, in the world outside Eastern Europe, the U.S.S.R., and China (whose reserves are unknown).

In areas with high logistic costs, small and lean deposits are obviously less economic or non-commercial, as in the North Sea, Arctic Canada, or central Australia. In many places small accumulations can, however, be economic. While only the probability of elephants will attract the high finance necessary to mount an elephant hunt, there is still good eating from a successful rabbit hunt in the right place. In other words, there is still much work to be done in finding even small sources that are strategically placed.

The People's Republic of China provides examples of this point. The chief coal deposits are in the north, away from the industrial centres. In order to diversify industrial location, the exploitation of small deposits, which could be worked by traditional, labour-intensive, indigenous methods, rather than by modern and costly machinery, was encouraged. This policy allowed the Chinese to gain self-sufficiency and maintain their self-reliance, without heavy imports of foreign machinery and expertise for hard currency. Built

10

with local resources and managed locally, such operations form, particularly south of the Yangtze River, a significant proportion of China's coal production. Coal, it is estimated, currently provides at least 80 per cent of China's total energy requirements. The Chinese similarly have many useful small hydro-electric schemes locally constructed and managed. Perhaps the lesson here is that, particularly in developing countries, the geologist should be encouraged to look for sites for many small hydro-electric schemes, rather than for the few sites for colossal schemes which may effect economies of scale, but which may also only bolster national pride at the cost of profound environmental changes for both land and people. While our knowledge of the formation and occurrence of coals, from anthracites to lignites and peats, is better than for the hydrocarbons, in many instances that knowledge is more apparent than real. There is still a need for continued exploration for coal. The possibility, even probability, of the existence of a rich coalfield around Selby in north Lincolnshire, England, may have been accepted by some for years, although it was proved only in 1972. Yet Britain is one of the best and most rigorously mapped countries in the world. The very large figures commonly quoted for world coal reserves are usually for original coal in place, not for remaining economically recoverable reserves. This is an important point which will be discussed more fully later, but Selby can illustrate another point.

Exploration effort varies with changes in world or national political and economic factors. The examples from the People's Republic of China illustrate how such forces influence the objectives of exploration. The Selby find resulted from a renewal of exploration effort because of possible changing world conditions. The decline in the coal industry, particularly in Europe, in the late 1950s and in the 1960s was due to the availability of plentiful, cheap oil, but some appreciated that this might not always be so. Hence the National Coal Board in Britain revived its exploration.

In the case of the uranium ores, there was intense but localized activity in the early 1950s. Then came a lull as problems arose in nuclear power production and as military requirements slackened. Exploration revived in the U.S.A. in the late 1960s, as America's growing dependence on imported oil became obvious. In late 1973 the insecurity resulting from that dependence was made strikingly plain when the Arab oil producers placed an embargo.

11

The energy resource industries

There are, however, differences between the nuclear industry, which is the youngest energy resource industry, and those established previously. The cost of the raw material is a small part of the cost of production of nuclear energy, for comparatively small quantities are required. The call for even these quantities will be reduced as the technical problems of breeder reactors are solved, perhaps towards the end of the century. Later, controlled fusion, using deuterium and lithium, may become the main method of exploiting nuclear energy. This forecast trend, together with the low cost element of the material, the small quantity required and, also, the ubiquity of low-grade deposits, make the nuclear physicists often play down the need for new sources. They tend to quote estimates, or rather guesstimates, of the total amount of uranium in the earth's crust, at say 10^{14} tonnes U, or state that the mining of 2×10^9 tonnes of granite with 4 ppm of uranium, burnt in a fast reactor, would supply the world's present annual total consumption. Thorium is even more plentiful than uranium. Present demand is only some 1000 tonnes per annum and, though it will be used in more gas-cooled reactors in time, known reserves of rich deposits are far in excess of forecast needs. Supplies of deuterium and lithium for fusion are almost limitless.

However, Bowie (1974) has suggested that, by 1980, production of uranium from known reserves will be at 60 000 tonnes per annum, which is about the medium-range forecast requirement for that year, and that to ensure continuity of supply thereafter more than twice the reserves found since the 1950s must be found at twice the speed. The reserves considered are those in the $20/kg category, and there is usually a minimum lead time of eight years, reckoned from discovery to production. Hence there is a need for new sources of uranium, particularly in the form of rich deposits. Indeed, more work is needed on the habit of occurrence of uranium in order to direct exploration most efficiently and economically.

Exploration for geothermal energy sources and their exploitation by electricity generation have been restricted in the past to the few of the more obvious manifestations around hot springs in Italy, New Zealand, the U.S.A., Mexico, Japan, the U.S.S.R., and Iceland. Hot endogenous fluids have also been utilized for direct heating in at least nine other countries. Since 1970 United Nations technical assistance missions on geothermal energy have been carried out in nineteen countries. The hot springs may be likened to the surface seepages of petroliferous areas; they prove the presence of the resource but the finding and tapping of a commercial reservoir, to convert the resource to a reserve, may not be simple. Geophysical techniques may help, as with petroleum, but only drilling can prove the reserves. In the western U.S.A. much of the exploration and exploitation in the late 1960s and early 1970s has been carried out by oil companies. They have brought to bear not only their drilling expertise but also their geological ideas. Again, as in the petroleum industry, the early exploration was not only confined to seepage areas but was also unsystematic. As knowledge increased, prospective non-seepage areas have been investigated, using all possible scientific methods. In Italy the discovery of steam fields in the Mt. Amiata region, about 70 km south-east of Lardarello, the cradle of the geothermal industry, was a break-through. There were no surface thermal indications and this dis-covery is acknowledged as a fundamental turning-point in research on geothermal fields. This illustration demonstrates the manner in which research in new areas can be improved by adopting and adapting advanced prospecting methods.

The total terrestrial heat flow is enormous and it has been said that if the total could be transformed into electricity by a process having a 10 per cent efficiency the resulting energy would cover the forecast world electricity demand up to the year 2000. As with all the other global figures, this may be interesting but is of no real significance. One must find the hyperthermal areas, the 'hot spots', where magmas are near the surface and have heated both the rocks and the fluids in them. These fluids may rise to the surface as steam, hot water, or hot brine or may be trapped by impermeable strata and so cause sealed convection currents. In the northern Mexican Gulf the sediments of the rapidly sinking basin in places contain fluids of abnormally high temperature and pressure, the result of geochemical hydrodynamic processes which are not fully under-stood but may be a potential energy source. In brief, there is plenty of scope for geological effort in developing geothermal energy.

In the geological section of the exploration team there are many places for others than the mountain climber. The palaeontologist, palynologist, petrologist, sedimentologist, structural geologist, and many other specialists may be part of the team for the period when their specialities have particular relevance. The composition of the team may vary during the development of exploration; and we have

13

already seen that the intensity and direction of the effort may vary. Hence the exploration geologist should be as broad-based and broad-minded as possible. One oil company in particular did not differentiate, as did most, between the explorer and the oilfield or resident geologist; all geologists had training in both aspects of their profession. This gave flexibility in staff management and also greater security to the individual geologist; otherwise there would be a tendency for the explorer to be the first to be declared redundant when economies had to be made. However, in the foreseeable future, as has been stressed, there should be a great and continuing need for exploration geologists in all the energy resource fields.

3. Production

The geologist also has a traditional role in the production phase which follows successful exploration, whether in the coalfield, oil or gas field, or ore mine. In this phase and in some locations the engineer may tend to overpower the geologist, sometimes just when geology properly applied to an extraction problem might well guide the engineer to a better solution. There are two parallel difficulties. The engineer, by choice and training, seeks precision and he prefers to build a machine, or devise a technique, to perform a standard task; such a machine or method can be precise and reliable. Moreover it will be less costly than one which needs built-in flexibility. The geologist cannot predict the characteristics of the rocks and their fluid contents with the same precision which can be applied to the design of simple machines or rigid techniques or practices. The infinite variety of nature requires that all machines trying to exploit natural resources must have a fundamental flexibility, and this is least costly if it is built in on the drawing-board. Unfortunately the engineer may not appreciate the impossibility of precise forecasting; and on the other hand the geologist may not appreciate the limits of the capabilities of the engineer's tools. In too many cases the problem is not discussed as a whole from the beginning; the engineer does not ask the right questions nor does the geologist offer the right information.

The need for all to work as an integrated and balanced team is as important in production as it is in exploration. Indeed, the interdependence of the scientist and engineer will become more obvious as the need for enhanced recovery becomes more apparent as deposits are progressively depleted. This area, that of increasing the recovery factor, is one to which much greater effort in both research and development of applied geology must be directed. The rewards in all forms of the non-renewable resources can be very great.

The petroleum industry can provide a graphic example. Paul Torrey (1963) has estimated that the percentages of the oil in place in known fields which were then expected to be recovered were as follows: U.S.A., 33·3 per cent; South America, 21·4 per cent; U.S.S.R., 53 per cent; the Middle East 38·4 per cent. Since then there has been considerable research and new techniques have been developed, but it is doubtful if many more points have been gained. Yet if the Middle East recovery factor could be raised by a further 10 per cent, the extra reserves recoverable would equal ten times the established proved reserves in U.K. waters estimated at the end of 1975. Even these British reserves could produce enough oil to make the U.K. self-sufficient through the early 1980s. Such an increase in the Middle East may be impossible, for many of the fields have been produced according to the best methods of oilfield practice for a long time. Furthermore the nature of the fields, particularly in Iran, is such that methods which have stimulated production elsewhere may not be applicable. The high factor in the U.S.S.R. is partly due to their practice of applying additional water-drive to their reservoirs from the beginning of production. This combines the two phases of primary and secondary production, which are often separate elsewhere, and in some fields has achieved 65 per cent of oil in place. This is another example of the influence on resource development of the variety in natural deposits.

Recovery in underground coal-mining has improved. New machines have raised productivity per man shift, but labour is still a problem and will continue to be until some form of unmanned, remotely controlled device, perhaps of the type invented by Professor M. W. Thring of Queen Mary College, London becomes common practice. When, and if, some such scheme becomes standard, there will be a very great challenge indeed. Thin seams might become economic and slight variations in the waste spoil would have to be instantly recognized and correctly interpreted to ensure proper supervision of the mole. This could be an even harder task than that of a resident geologist on a 'wildcat' oil well who has to guide the driller by monitoring the rock cuttings.

The geologist has an important part to play in the production phase of a geothermal field. No field has the simplicity which the engineer deems ideal. A lot of work has to be done to understand a local fracture system; indeed, there are probably many problems of geological engineering or engineering geology which have not yet been tackled with all the modern knowledge and techniques which they merit.

On the production side of all the energy resource industries the geologist must be more conscious of cost factors than he may possibly have been in the past. It is no longer reasonable for him to ignore the economic parameters of his own industry. Forty years ago it was not uncommon for a geologist employed in an oil company to stress that he was a geologist, not a petroleum geologist. This attitude may have fulfilled an egocentric need but it was not helpful to his work. Such 'ivory tower' attitudes are diminishing; and so they must if the geologist is to be fully accepted as a professional man in industry. Geology is also becoming more numerate. Perhaps it was partly the lack of numeracy which in the past encouraged the growth of splinter groups of those who considered themselves more attuned to current modes. Many were closer to the popular conception of a scientist. Certainly geology is commonly considered to be only one of the earth sciences, whereas, by definition, geology encompasses all knowledge of the earth.

Be that as it may, there is no gainsaying that in all the resource industries the applied geologist must be conscious of the economic restraints implicit in his work. Indeed, as is more obvious in the less traditional roles which the geologist can play, he must also be conscious of the politico-economic and social implications of his work.

4. Reserves

Exploration is aimed at discovering resources, which are the natural stores of future energy supply; reserves are the economically recoverable part of those resources; production is the utilization of those reserves. In all the non-renewable energy sources there is primarily a finite *resource base* of which part has been produced, part has been measured as being the *proved reserves* (which can be produced under current economic and operating conditions), and further parts that are less well known and may be termed *possible*, *probable*, or *remaining reserves*, according to the degree of certainty. In Fig. 2.1 the components of the resource base are illus-

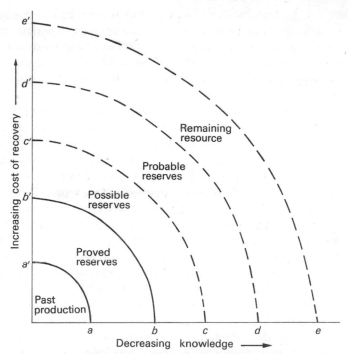

Fig. 2.1. Components of the resource base.

trated on the two criteria of geological knowledge and estimated recovery cost. Our knowledge of the earth is sufficient for us to know that there are finite limits to the coals in place, hydrocarbons in place (whether in solid, semi-liquid, liquid, or gaseous states), uranium, and thorium. Our knowledge is insufficient to know what the numbers are and we can only make guesses, usually by analogy, making comparisons with the best-known and the least-known areas.

While this simple model is basically true for all non-renewable resources, practice within the different industries has led to various concepts of reserves. This has meant that up to the present the figures for reserves are not comparable, for the words used have different meanings. This is a most important point, particularly because, as will be emphasized later, an increasing number of ill-informed people are starting to use world energy data. The differences in reserves concepts can be illustrated by variations on Fig. 2.1.

The energy resource industries

The petroleum industry practice fits closest to the basic concepts. *Crude oil proved reserves* are the estimated quantities remaining in the ground which geological and engineering information indicate, with reasonable certainty, can be recovered in the future from known reservoirs under current economic conditions, using current operating techniques. Variations in interpretation of the term 'with reasonable certainty' alter the position of point b on Fig. 2.1, and relative differences, say, in location to market and other factors, also vary the position of b'. When, about 1950, offshore sedimentary basins could be added to onshore basins as prospective areas, the resource base of the total oil in place was not altered. Our estimate of its size was, however, expanded. More importantly, first the line dd', then the line cc', then, as discoveries were made, the line bb' moved to the right. Now, in its turn, offshore oil is helping to move the line aa' to the right. In late 1973, when crude oil prices moved dramatically upwards, point b' could be moved upwards. The amount of proved reserves was increased, theoretically, as the line bb' moved closer to bc', but as yet no one knows how close.

Consideration of natural gas reserves is more complicated than that for crude oil. Gas which is associated with oil must be treated differently from that which occurs in a non-associated form. In the first case, whether the gas is in solution or in a gas cap, it is an essential part of the oil-production mechanism. It cannot be produced except with the oil until all or most of the recoverable oil has been produced. Hence the gas resource, though measurable, is in a different category from gas in a gas field, not associated with oil and freely available. Another variant is that in Canada proved reserves of natural gas are reported on a marketable and not a recoverable basis; marketable or sales gas means a mixture of methane originating from the raw gas, if necessary after processing, which meets specifications for use as a domestic, commercial, or industrial fuel or as an industrial raw material. Natural gas liquids (N.G.L.), such as condensate, natural gasoline, or wellhead gasoline, are the hydrocarbons which in the reservoir are in gaseous form but become liquid in the well or at the wellhead with the lowering of temperature and pressure. Most countries do not estimate N.G.L. separately but include them with the crude oil.

Solid hydrocarbons, in organic-rich shales and as tar sands, have been found in many countries but in few countries have they been adequately measured and assayed. In oil-shales the hydrocarbons are in the form of solid kerogen, which has to be heated and 'cracked' before it is of practical use. Shales have been exploited in Scotland and Sweden and are now being exploited in Estonia and the People's Republic of China. Much research work has been done on the rich shales of Utah, Wyoming, and Colorado, and pilot plants have been constructed in connection with them. With the current increased drive for self-sufficiency in energy sources in the U.S.A., there is much activity but as yet no single proved commercial process. Hence all estimates of reserves of oil shales, outside the deposits being produced, must be considered in the probable or possible reserves categories and are not proved reserves.

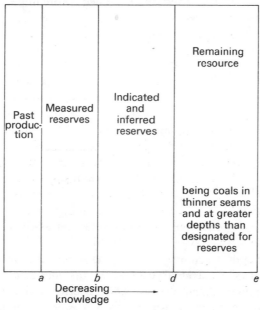

Fig. 2.2. Traditional coal industry reserves concepts.

The coal industry traditionally defines its *measured reserves* as those quantities measured on reliable data as being at less than 1200 m below the surface and in seams not less than 30 cm thick. No economic parameter is involved, some using the figure of original amounts and without subtracting the quantities already produced. *Indicated* and *inferred reserves* are the additional quantities, within the same stated limits of depth from surface and thickness of seam, which can, from incomplete investigation or from relation to

19

measured reserves, be reasonably assumed to exist but of which only approximate estimates are possible. The traditional coal industry concept might be illustrated as the variant on the basic concept shown in Fig. 2.2. In recent years, however, there has been an increasing tendency to introduce an economic limit; this has markedly reduced the commonly quoted measured reserves figures.

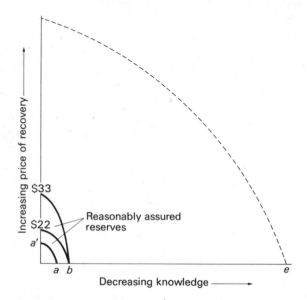

Fig. 2.3. Nuclear energy reserves concepts.

The nuclear industry makes cost of recovery, or rather price of recovery (since an element of profit is included), a criterion of reserve definition. *Nuclear reasonably assured resources* are normally quoted as those amounts which can be profitably recovered at a price up to $ 22/kg U_3O_8 or, in a second category, at between $ 22 and $33/kg U_3O_8. The joint Nuclear Energy Agency–International Atomic Energy Agency annual report has abandoned a third, higher price category. Broadly, therefore, Fig. 2.3 would represent the nuclear industry reserves concept.

Estimates of recoverable 'synthetic' oil from the Athabasca Oil Sands, Canada, of Lower Cretaceous age, might illustrate the important differences which different concepts can make to estimates of reserves; Table 2.1 gives some illustrative figures.

TABLE 2.1
Athabasca Oil Sands: Estimates of reserves and resource base

	Quantity (U.S. barrels × 10⁹)
1960. Original estimate of the 'synthetic' oil recoverable on the Mildred Lake Lease	0·6
1968. Official proved reserves for the existing plant, (cf. strict proved concept for North American oil)	6.0
Estimate of recoverable oil by opencast mining methods (cf. nuclear energy concept with known plant costs)	38·0
Estimate of recoverable oil by opencast mining and thermal methods *in situ*, not yet proved (cf. coal concept)	300·0
Oil-in-place estimate for the whole area (cf. resource base)	600·0

Let us now consider how reserve/resource figures originate and are compiled and so gauge their validity. The best-established figures are those compiled with the help of the geologist directly involved in the site, whether coal-mine, oilfield, ore mine, or geothermal field. Only local knowledge will provide a basis for recognizing how well-grounded are the assumptions used in the calculations. These figures then go up the chain of command, whether company or ministry, and are liable to amendment by pride and/or prejudice, by rounding up or down on aggregation with other estimates made, perhaps, on other assumptions, all of which are gradually forgotten as the figures move away from their origin. The accuracy of all energy reserve and resource figures can therefore be considered only as 'fair'. For instance, any geologist can appreciate that the precision of the measurement of coal seams down to 30 cm thickness is more apparent than real. Yet figures with that apparent accuracy, giving an air of verisimilitude, may be in error by orders of magnitude and be most misleading to the planners.

5. Planning

Plans for many purposes by many people, inside and outside the energy industries, are dependent on energy resource data. The geologist is traditionally employed on the supply side of the industry, but his skills may also be of use on the planning side.

The unequal distribution of minerals in the earth's crust is an obvious fact to the geologist, but to some others it may appear to be the outcome of a malevolent conspiracy. Some years ago the reported occurrence of specks of a substance resembling bitumen in the drill cuttings of an exploratory well was seized on as incontrovertible evidence that a large oilfield had been found but its existence was being concealed by the exploring company. The success ratio in the northern North Sea has been well above average for the world. If success is not as good in other parts of the shelf around Britain, some may well argue either that this is because the operators did not wish to find more or that it is all the fault of perfidious Albion!

The geologist's knowledge of the infinite variety of nature should give him a balanced view of such a position. He can probably tell stories from his own experience of that variety and the luck of the game. Two explorers may be sent to two different areas. One recommends drilling and a field is discovered. The other recommends no further work. Professionally the work of the second man might be of far higher calibre and his acumen far greater than that of the first. The custom in the old Anglo-Persian Oil Company was that the only awards made on a discovery were that the geologist, the drillers, and the engineers who were on the well when it 'came in' got gold watches; the geologist could easily not be the one who recommended the location. That wise practice discounted luck, let merit show in the quality of work, and acknowledged the complexities of nature.

In the interpretation of production data there are three common fallacies on which a scientific viewpoint is very necessary. First there is the problem caused by aggregating, which is applicable to all resource fields but well illustrated by the contrast between the smooth curve of the aggregated annual world crude oil production and the jagged curves of some of the components, as shown in Fig. 2.4. Secondly there are the misleading uses of reserves/production ratios. These ratios indicate the number of years that the specified reserves would last if production were to continue at the specified rate. The ratio is then regarded as the number of years that the reserves will last. Then comes the comparison with other sources, ignoring the great differences that exist between different types of reserves, which we have already stressed. A further illustration can be taken from the nuclear field to emphasize that the method of conversion of the reserve into production can be important. There is a 30 : 1 difference in the energy value of uranium whether it is used in a fast reactor or a thermal reactor. Thirdly there is the misuse of

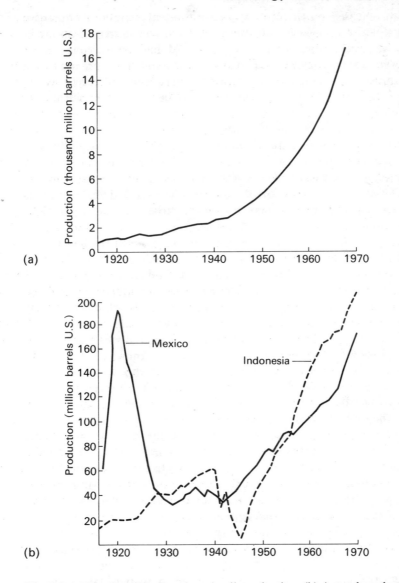

Fig. 2.4. (a) Total world annual crude oil production; (b) Annual crude oil productions of Mexico and Indonesia.

data released on the discovery of a mineral deposit. Responsible companies or agencies in mining and petroleum are cautious in their releases but, no matter how cautious they are, the mass communication media quickly summarize and headline, omit vital parameters, and ignore assumptions. Here are three areas as examples where the geologist should be fitted for a wise educational role.

On the reserves side in particular the geologist has perhaps a greater than average trained ability to form, almost instinctively, a sound evaluation of data as presented. To the geologist an ore body is not just a weight of x tonnes, but has a shape which he immediately tries to visualize. A coal seam has lateral extent as well as thickness and may or may not be disturbed by many faults or pinch-outs. An oil reservoir is not a cavern, but rock, with porosity and permeability and various characteristics, whether limestone or sandstone, shaly limestone or coarse sand; it is these that will govern, in part, the recovery factor and production rate.

The role of the industrial geologist in the planning team must be neither overplayed nor ignored by default; in the resource planning field he may well have a most important role.

6. Laws and regulations

There is a well-established role for the geologist as an expert witness or 'certifying witness' in matters bordering on the legal. This is particularly so in countries where the mineral rights are held by the individual surface rights holder. There are many consulting geologists who appraise properties before development financing is attempted. Many banks, particularly in the U.S.A. and Canada, employ geologists to advise on energy matters and not only on the evaluation of properties as collateral for loans; more will probably do so in the future. There are other consultants who stress their expertise in exploration or production. They may be used as impartial advisers to regulatory bodies on new or revised regulations, as independent arbitrators, or even as watchdogs to ensure that licensed explorers or exploiters are keeping to the terms of their licences. In some countries the geologist must be certified as a professional geologist before he can practise. These are only some of the opportunities for independent consulting geologists in the energy resource industries.

Increasingly, however, as energy resource industries gain national identities, and the roles of the international companies change, as they did most dramatically in the petroleum industry in October 1973, governments will become more intimately involved in the energy resource industries. Government departments throughout the world will grow as they assume more direct energy planning and regulation of production, conservation, conversion, and utilization of all energy resources. The problem, particularly in Europe in 1974, was that the call on government departments for such services showed up a lack of trained personnel. This lack should be only a temporary one, and it should mean that more energy resource geologists will be employed in government service.

One new area of interest for the industrial geologist is that of the jurisdiction of the deep sea bed. The Geneva Convention of 1958, ratified in 1964, defined the Continental Shelf in juridical terms as 'the sea bed ... adjacent to the coast but outside the area of the territorial sea, to a depth of 200 m, or, beyond that limit, to where the depth of the superjacent waters admits of the exploitation of the natural resources of the said areas.'

Recent years have seen the exploitation of petroleum from ever-deeper waters and exploration in even deeper waters still. There has also been a growing possibility of exploiting other minerals from very deep waters. Hence there was increasing general concern that the developed industrial countries, alone capable of such exploitation, would gain all the amazing wealth waiting to be plucked from the ocean floors. The United Nations started discussions and the first task was to destroy the myth of immediate and enormous wealth. Almost all nations thought that the deep ocean floor should be developed for the good of all mankind. The problem was as to where the juridical continental shelf should end and the juridical ocean floor begin, or whether there should be some form of a buffer zone. Geologists have taken an active part in the discussions and have been asked to delineate potential areas of commercial interest. They have also been asked to delineate the edge and foot of the continental shelf in the geological sense, as possible natural topographical boundaries, which many consider to be the best political boundaries. In any demarcation of the oceans the status of islands, whether of continental or oceanic type, becomes a problem (hence the importance to Britain of the continental nature of Rockall). The interest of the petroleum geologist in the subject of jurisdiction of the sea bed will be obvious. A conference on this subject was held in

Caracas in June–August 1974 under United Nations auspices, but no agreement was reached and the discussions, in which geologists have taken part, have continued.

Geologists also take part in the United Nations Technical Assistance Missions to developing countries. These are usually concerned with ascertaining the facts, helping to plan the best use of energy resources, and establishing regulations to control their exploitation.

7. The environment

Public concern with pollution and the preservation or restitution of the environment has emphasized a number of aspects of the geologist's work in the energy resource industries, especially as it has been paralleled by concern for conservation in the sense of eliminating waste in the utilization of all natural resources.

Only in a few highly specialized and highly sensitive areas does the exploratory phase of the energy resource industries present any significant environmental problems. In most areas geological work on a reconnaissance or detailed scale has negligible effect. The geological hammer can cause only local damage. Augers, bore-holes, drilled wells, and some geophysical surveys may cause more disturbance but with the care which is normally taken, and is obligatory in many countries, very little permanent harm is done.

Permafrost areas of tundra are peculiarly sensitive areas environmentally. They have gained public prominence because of the planned production of crude oil from the North Slope of Alaska. Ill-considered, over-hasty action by reconnaissance seismic crews and in the first drilling operations alerted the U.S. Government and the American public to the considerable danger of damage to the environment if due care was not taken. Permafrost, affecting the earth in some places to a depth of 600 m or more, is a geological phenomenon which has been comparatively little studied, even though from 1940 onwards there has been considerable activity in the Arctic on military installations and Eskimo resettlement. Since the discovery of oil at Prudhoe Bay in 1968 this situation has changed but much work still remains for a team in which the skills of the geologist, the botanist, the soils engineer, and the hydrologist all have their place.

An interesting application of ERTS imagery is in the examination of the area around Umiat on the North Slope of Alaska where there had been extensive geological and geophysical activity and some test drilling immediately after the Second World War. ERTS revealed no large-scale environmental degradation and most of the trails had been revegetated. Even where erosional features had been caused by the breaking of the tundra surface these areas had also been revegetated. This ERTS scan showed that the tundra can recover, at least from some of the man-made changes, in 25 years, instead of the centuries which some had said would be required for restoration. This topic also illustrates that there still are little-known areas of the world into which the energy resource geologist goes.

In the production phase there is a greater potential for damage and opencast coal-mining operations are now required to include restitution of the topography, topsoil, and revegetation. In the U.S.A., the ERTS has been used for periodic monitoring of opencast operations. In the opencast mining of the Athabasca Oil Sands in the almost unpopulated McMurray area of north-east Alberta, the cleaned sands have to be handled so that the environmental balance is not disturbed. The disposal of the dust-like spent shale from the proposed plants in the oil-shale area of the Piceance Basin of Wyoming, Utah, and Colorado is a very difficult problem, for water for wetting and binding is very scarce. In any area large spoil dumps from deep mining or disposal lagoons can be an eyesore but sometimes geological advice can help. Restoration of the landscape is now accepted as part of the total operation and some even claim that beneficiation of the environment can result.

In the petroleum industry there is a minimum of disturbance in the production phase, except potentially from pipelines and at sea. Techniques are now well developed for eliminating, except in the rarest of circumstances, the blow-out of drilling wells with oil spewing over the landscape or seascape. Production in Iran started in 1908 and the first well to go wild was a gas well in 1951; this was under control within days. In the twenty years of offshore production there have been very few disasters. There is obviously a potential risk, of which the production crew is well aware; and the geologist has special responsibility in recognizing marker beds during drilling which may assist in predicting blow-outs.

Another potential hazard during drilling can be illustrated from Malta, though the operation was exploration and not production. This was a most delicate task, for there was the danger of potential damage to the drinking-water supply for the whole island. The potable water reservoir is a freshwater lens, floating on the almost horizontal saline water surface at sea-level, right across the island in the highly permeable, homogeneous, soft limestone of which much of the island is made. Drilling with air as lubricating medium, instead of mud, and then casing off the freshwater zone solved the problem.

In the matter of pipelines, whether on land or on the sea-bed, the geologist should play an important part. First there is the variation in potential risk to the environment with the type of rock over which the line runs. A leak in a coal-slurry pipeline or, more critically, in an oil line into chalk or coarse sandstone obviously has a far greater chance of contaminating groundwater than a spill in clay. The geologist can delineate these areas and so can advise where greater safeguards should be taken. It would be a waste of effort and money to take the same measures over the clay as over the chalk, but it would be irresponsible to take the same safety measures over chalk as over clay.

Mention has already been made of the highly sensitive permafrost environment. The Alyeska Pipeline from Prudhoe Bay across Alaska to Valdez, on which work started in the summer of 1974, will be the first oil line across permafrost and will probably take three years to lay. As noted above, considerable research was done and improved plans eventually received sanction on the advice of the U.S. Geological Survey, the statutory environmental watchdog.

Offshore pipelines are less liable to accidental damage from trailing anchors and other hazards if they are buried. Geological advice should always be taken on the nature of the sea-bed, as it was before the Pipeline Under The Ocean (PLUTO) was laid in 1944 to provide the Allied Armies with petroleum products across the English Channel. Nowadays sampling methods are better and instruments, such as sideways scanning sonar, are available to assist the geologist, yet there have been recent cases where inadequate use of geological advice has caused difficulties which could have been avoided.

In the nuclear energy industry the environment can be protected during ore mining. Transport presents no geological problem, and the ultimate site for the power plant need pose no greater problem than for any other major civil engineering construction. The geologist may, however, be involved in the safe disposal of radioactive wastes from thermal reactors. In the field of geothermal energy a collapsed well, with a periodic mud fountain, may even become an added tourist attraction rather than an eyesore.

In the hydro-electricity industry the geologist obviously has an important role in exploration and the choice of optimum dam and reservoir sites offering maximum safety. He also has a vital role when tunnelling is entailed, particularly when, as in the Snowy Mountain scheme in south-east Australia, the water is to be taken under the mountains to another river basin to yield more power and also to provide for irrigation; there are many other such dual schemes for irrigation and power. The value of small schemes has been mentioned above, yet in some places reservoirs are said to be alien features. The geologist can do little about such a broad issue, but he may well be able to predict the effect of drawdown and erosion by wave action at the edges. He may also be able to devise landscape features to mask what might otherwise be offensive strips of bare soil or decaying vegetation.

Little can be done by the geologist to reduce the visual effect of the surface storage of coal. In some cases, however, the obtrusion of oil storage tanks on the landscape can be avoided by the substitution of underground storage. The geologist can advise on the worth of old coal or other mines, or on the excavation of caverns in salt or in rock. The Federal Republic of Germany is using salt caverns for its strategic storage. Sweden has long used storage in caverns made in igneous rock and abandoned room-and-pillar coal workings are used in South Africa for similar purposes.

Finally, the geologist has an interest in the whole of the environmental field which is surpassed by none. His interest is in his knowledge of the earth and this does not limit him to rocks: a geologist is more than a 'rock hound'. On land, rocks disintegrate into soils with characteristics which vary with the parent rock and which support specific types of vegetation. That fact can significantly guide geological interpretation of aerial photographs. The exploration geologist working on the ground learns to understand the jungle and desert as well as the countryside nearer home. The geologist, by inclination and education, is a trained observer and wanderer over the earth. As such he learns about the present-day environment better than most others. He also has to know about other, older environments, whose 'stories are in the stones'. This fosters an appreciation of how environments have changed over geological time and will continue to change. The environment which encouraged the evolution of the dinosaur in western Canada was located in a region where later millions of buffalo thrived and now millions of bushels of wheat are grown. This appreciation of change helps to balance the more extreme conservationist arguments (which were more commonly heard in 1966 than in 1976) for a return to the 'natural environment'. A general background and style of thought are common to all geologists and are not restricted to those engaged in the energy resource industries; but these industries are very often the target of criticism, and sometimes even abuse, by the public. It is thus relevant to consider the geologist in this particular field.

8. Society

In this discussion of the energy resource aspect of industrial geology, we have considered the part which the geologist plays in his traditional roles in exploration, production, and reserves evaluation, extending into the less familiar roles concerned with planning, laws and regulations, and the environment. We have seen that energy data are complex and that there is a growing awareness of their importance in daily life. This awareness is not restricted to the industrially developed countries but is increasingly felt in the developing countries. The point has also been made, and it is equally true for all countries, that the public and many politicians and civil servants are being faced with complex data that few people can interpret adequately. Since energy is so important to society, let us sketch four types of growing interdependence in the energy resource field which may significantly affect society in the future: the interdependence among the energy industries, among the producers, among the consumers, and among the producers and consumers.

The interdependence among the energy resource industries was illustrated by the argument at the 1967 Annual Meeting of the American Association of Petroleum Geologists that members should consider themselves 'energy geologists' and not just 'petroleum geologists'. The keynote speech was given by the geologist chairman of an oil company which had already a large uranium-producing department. Oil companies own the second, third, and tenth largest of the ten coal companies which currently produce 50 per cent of the coal mined in the U.S.A. The National Coal Board in Britain is exchanging research data with the U.S.A. and is spending substantial sums on research into the conversion of coal into liquid and gaseous fuels. Coal, once the source of manufactured town's gas, was ousted by natural gas and research is now being done to return to this same use. It was calculated that gas made from coal in the U.K. would cost, at mid-1973 coal prices, about 7p per therm. This is about three to four times as expensive as beach prices of North Sea Gas, but only 50 per cent more than current industrial gas prices. British Petroleum is actively seeking coal reserves world-wide. The British Gas Corporation was using light petroleum fractions to make town's gas, in a plant integrated with an oil refinery in Kent, and was importing liquefied natural gas from Algeria before the discovery of natural gas in the North Sea. A number of oil companies have taken interests in the nuclear field and, as stated above, oil companies in the U.S.A are actively exploiting geothermal energy.

These are the outward signs of interdependence of the energy industries but there seems every possibility that there will be more integrated planning in the future than in the past. There are increasing pressures towards conservation of energy, optimum utilization of natural resources, optimum efficiency in extraction, conversion, and use, especially of the fossil fuels, and restraints on some practices for reasons of environmental conservation. The peculiar characteristics of each energy resource will have to be applied to the greatest advantage. Hence there will be continuing pressure for industrial geologists in the energy resource field to be specialists in one aspect but informed generalists in all.

The interdependence among producers' is epitomized in the Organization of Petroleum Exporting Countries (OPEC) and the kindred body restricted to Arab countries. Their power was dramatically illustrated in October 1973 by their joint political action in regulating the destination of petroleum being exported from their countries according to their own assessment of the friendliness of customers to the Arab side in their war with Israel. There is little doubt that demand for petroleum in future will have to be tailored to supply and not the reverse as was the case during the recent past. Furthermore, producers will control supply according to their needs as they see them. This attitude may become common to all energy-resource producers; and though petroleum is now the dominant energy source in international trade, coal may return as an important component.

The interdependence of the petroleum-consuming countries became manifest during the 1973–4 winter shortages. Collaboration between these countries is less obvious, but may well be necessary before the real interdependence between producers and consumers, most obvious in petroleum, but perhaps later in all fields, can bear practical fruit to mutual advantage.

Since October 1973, it can be said with some justification that in world energy, not only on the resource side, there is a new game, there are some new players, and new rules are being written as play proceeds. The geologist in the energy resource industries can have an important part all over the field.

References

BOWIE, S. H. U. (1974). Natural sources of nuclear fuels. In Energy in the 1980s, *Phil. Trans. R. Soc.* **A276**, 495–505.

TORREY, P. D. (1963). World oil reserves. *Proc. 6th Wld Petrol. Congr.* **2**.

3. Oil and natural gas

1. Introduction

Until relatively recently the conduct of exploration and production activities by the oil industry has had little impact either on the general public or on academic geology outside the United States of America and a few of the other major oil-producing countries. America was historically the centre of most oil activity; thousands of geologists have been employed there by the oil and gas companies, and academic institutions have been closely involved in the special training and research associated with oil and gas geology. In most other countries, and in Britain in particular, the lack of contact with activities in remote countries and the absence of initiative for research has not encouraged activities connected with the exploration for and production of petroleum in the universities and training centres. The discovery of large-scale reserves of hydrocarbons in some European countries has suddenly aroused an interest in oil activities. It has also emphasized the unfortunate lack of indigenous trained staff. This is a phase that even the OPEC countries have gone through, and their recent programmes of increased production have still left most of them lacking trained staff.

It is clearly impossible within a few pages to give a detailed insight into the everyday activities of oil geologists; this brief review can give only a distillation, based on the writer's experience in the oil industry, of the chief aspects of geology involved in the exploration and production of oil and gas, the work done by oil geologists, and the prospects for the future.

2. The nature and significance of oil in the subsurface fluid regime

Apart from those concerned in producing water or oil and gas, not many geologists may give much thought to the fact that a significant proportion of the upper crust of the earth consists of fluid, mostly water. Nor do they give much thought to the properties of porosity and permeability whereby rocks can contain fluids, allow their movement, and lead to the production of water and hydrocarbons at the earth's surface.

Most sedimentary rocks are porous unless metamorphosed. The porosity of nearly all sediments is in the range 5 to 50 per cent of the total rock volume. This represents a vast volume of fluid-filled space, and an understanding of the origin and nature of porosity and permeability and their effect on the movement of underground fluids is a matter of great concern to the oil geologist.

Porosity exists in all sedimentary rocks from the moment of their initial deposition. During the course of geological time it is modified by various processes which mostly work towards its diminution. The processes affecting porosity are numerous and they can be syngenetic and diagenetic. They include compaction under the weight of subsequently deposited sediment; solution and recrystallization of the grains of the sediment by moving water and changes of temperature and pressure; reactions within the groundwater, which result in the precipitation of solid material can block the pore spaces. Heat resulting from deep burial, regional metamorphism, or from near-by igneous intrusion can partially or entirely fuse the rock and destroy porosity.

In clastic rocks, which are composed of redeposited fragments of particles of other rocks and their degradation products, the nature of the porosity and the permeability dependent on this porosity are related to both the size and the sorting of the particles. Mudstones and shales, which are composed of uniform small particles, have a high proportion of their volume composed of pore spaces but these are of very small size and the permeabilities of such rocks are low. A sandstone or conglomerate may have very large particles indeed and some large voids but may have a low total volume of pores and poor permeability owing to the presence of much fine material. The best clastic rocks for oil and gas reservoirs are clean sands of uniform and adequate grain size.

The carbonates form another family of rocks that have the necessary qualities of porosity and permeability to make them important oil and gas reservoirs. The carbonate reservoirs are primarily limestone and dolomite; their porosity is extremely varied in origin and type. Many carbonates are of clastic origin, i.e. they were deposited as fragments of shell or coral debris; others are chemical deposits of very fine grain size. Owing to the relative chemical instability of calcium and magnesium carbonates in the aqueous regime in which they are both deposited and subsequently exist, they are particularly prone to chemical change, recrystallization etc., with consequent changes of porosity and permeability. The process of dolomitization, which results in volume reduction, normally ensures that dolomites have good porosity and make good oil and gas reservoirs.

Another aspect of carbonate reservoirs is their much greater propensity for fracturing. Fractures commonly play an important role in the performance of oil and gas reservoirs. The movement of fluid through a reservoir rock, and hence the productivity of wells, depends on the matrix permeability, a function of porosity and pore-size distribution, and also on transmission along fractures. Loosely cemented and uncemented sandstones do not develop effective fracture permeability. Cemented sandstones and most limestones develop extensive fracturing and can also be induced to fracture by application of pressure by fluid injection. Many carbonate reservoirs, including the prolific Asmari Limestone of Iran, have low porosities and matrix permeabilities but are excellent reservoirs because of extensive fracturing.

The development of fracturing is a function of deformation by folding or by regional joint-forming stresses. The prediction of optimum fracturing positions, an important consideration in the location of, and production from, wells, calls for an understanding of the tectonic history of oil- and gas-producing structures.

The pore spaces of rocks are filled in the upper parts, above the water table, by a mixture of air and water. Below the water table the pore spaces are filled mostly with water. Down to a depth of a few hundred feet, sometimes to many thousand feet, the pore fluid is fresh water. At greater depth the water invariably becomes salt and this formation water, whose composition is an important consideration in drilling wells and producing hydrocarbons, is commonly highly saline and approaches saturated brine in composition. Such formation waters affect drilling muds and have an important influence on water-injection processes.

After water the next most abundant fluid constituent of the underground pore-space is the family of hydrocarbons. These hydrocarbons are the oil and gas with which we are here concerned.

Oil and natural gas

Hydrocarbons are not rare underground; they are of widespread occurrence in rocks of all ages. It is only concentrations of oil and gas in quantities that justify commercial extraction that are relatively uncommon. A knowledge of the porosity of rocks is an essential element in estimating the oil in place in an oilfield. Examination of cores and downhole logging techniques can be used to estimate the percentage of the reservoir rock that is occupied by pore space and the proportions of oil and water that occupy that space; virtually all oil-producing reservoirs contain some water, the range being normally within 10 to 50 per cent of the total fluid. From a knowledge of the total rock volume within the oilfield, derived from drilling information, geophysics, and other sources of data, we can use information on the percentage of this rock volume that is pore space and the part of this that is oil to estimate the total volume of oil in place in the reservoir.

The oil in place can be estimated with an accuracy of about plus or minus 10 per cent. To obtain an estimate of oil recovery is somewhat more difficult and involves consideration of porosity, permeability, pressure, gas content, oil viscosity, water drive from outside the oilfield, and the production methods to be used, bringing in considerations of time and cost. The range of oil recovery in commercial oilfields is from 5 to 50 per cent of the oil in place. Recovery from good fields with reservoirs of high porosity and permeability is usually in the range 30–40 per cent of the oil in place in a period of ten to twenty years. Gas recoveries are much higher and recoveries of up to 80 or 90 per cent of the original gas in place are not uncommon.

A simple diagrammatic representation of the relative volumes of rocks and their fluid contents is shown in Fig. 3.1. This shows the relationship, represented by area, of the sediments and fluid contents in a typical oil-producing basin. The rock volume is the gross rock volume of the entire basin, i.e. of rock and its fluid-filled pore space. The inner, smaller circles show the relative volumes of the fluids filling the pore spaces.

The hydrocarbons with which we are concerned consist of an almost infinite number of combinations of members of a few main families of compounds. These families contain many separate compounds and isomers which have similar properties and are therefore difficult to separate and identify. The four main families or series of hydrocarbons that constitute the bulk of most crude oils are, in order of relative abundance, the normal paraffins (or alkanes), the isoparaffins, the naphthenes, and the aromatics (or benzenes).

The normal paraffins are saturated, straight-chain compounds of a homologous series having the general formula C_nH_{2n+2}. The simplest member is methane, CH_4. Methane and the three next higher members of the series, ethane, propane, and butane (C_4H_{10}) are gases at normal atmospheric temperatures and pressures and are the main inflammable constituents of natural gases. The higher

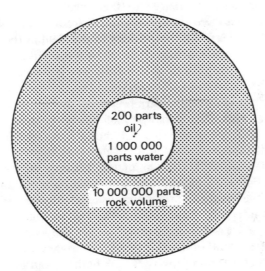

Fig. 3.1. Relative volumes of rock, water, and oil in a typical oil-producing basin. (By permission of the Royal Geographical Society.)

members of the alkane family, i.e. those with 20 or more carbon atoms, are waxy solids at normal temperatures. The naphthenes are saturated hydrocarbons with closed ring structures of the general formula C_nH_{2n}. The properties are similar to those of the normal paraffins, and they range from gases to solids. The aromatics are cyclic, unsaturated compounds with general formula C_nH_{2n-6}. The simplest and most common member of the series is benzene, C_6H_6. Aromatics rarely form the major constituents of naturally occurring crude oils.

This is not the place to go into the complexities of hydrocarbon chemistry but the potential can be appreciated when it is seen that it is mathematically possible for a paraffin with 18 carbon atoms and 38 hydrogen atoms to have 60 523 possible isomers.

There are numerous minor constituents of oils, including sulphur and nitrogen. Sulphur may form up to 5 per cent by weight of some crudes. Natural gases occur with up to 90 per cent by volume of nitrogen. There are also enormous numbers of metallic and non-metallic compounds as very minor constituents of most petroleums.

There is a voluminous literature on the origin of crude oil but not much firm information. Oils are mobile fluids and it is impossible to relate subsurface occurrences to their places and conditions of origin. Although there are a few who adhere to a 'planetary' origin for petroleum, nearly all the workers in this field in the oil and gas industry are convinced that it is derived from the diagenesis of organic debris trapped in sediments. Methane certainly is associated with some volcanic gases and is, of course, formed in abundance from thermal degradation of coal, lignite, and other organic materials. The variability of oils from different areas is illustrated in Table 3.1. Table 3.2 shows the variations that can exist in different reservoirs in the same structure. In the Iranian example there is a 2000-ft shale interval between the two reservoirs but the examples from Abu Dhabi are within a continuous limestone unit, the Thamama; the separate accumulations are related to permeable intervals. Table 3.3 shows the main components of various natural gases from various parts of the world. It is easy to tell by the content of hydrocarbons of higher carbon number, which gases come from provinces with both oil and gas rather than from dry gas areas.

There are two other aspects of the underground regime which are of great importance and interest to oil geologists: temperature and pressure. Both these physical characteristics have considerable bearing on the generation of oil and gas and its migration and trapping in reservoirs. They are also of importance in the practical aspects of drilling wells and producing fields.

Consider temperature: the average geothermal gradient is about $2\,°F$ or $1\cdot1\,°C$ for every 100 ft (30 m) of depth. Thus, on average a 10 000-ft well is hotter at the bottom than the boiling-point of water at atmospheric pressure. The critical temperature of water is $365\,°C$ and this will be found on average below 30 000 ft, a depth reached by modern deep wells.

There is considerable deviation from the normal geothermal

TABLE 3.1
Analyses of crude oils from various commercial fields

	Field			
	Forties (North Sea)	Boscan (Venezuela)	Oficina (Venezuela)	Burgan (Kuwait)
Specific gravity at 15·60 °C (60 °F)	0·842	1·000	0·856	0·856
API gravity	36·6	20·1	33·8	31·7
Sulphur content (% wt)	0·28	5·5	0·77	2·6
Asphaltenes (% wt)	0·55	12·0	1·8	0·7
Pour point (°F)	30	60	−5	−25
Wax content (% wt)	8·5	2·3	6·3	5·2
Viscosity at 38 °C (cSt)	4·48	19 400	4·7	8·8
Vanadium content (ppm)	<3	1200	39	25
Nickel content (ppm)	2			
Distillation products				
Gasoline (% wt)	19·0	1·4	19·2	15·0
Kerosene (% wt)	13·7	3·3	13·9	11·7
Gas oil (% wt)	21·4	10·0	21·4	16·5
Residue (% wt)	44·7	85·3	44·7	55·8

TABLE 3.2
Analyses of crude oils from different
production intervals in the same field

Field	Zakum (Abu Dhabi)		Ahwaz (Iran)	
Zone	Upper (Thamama)	Lower (Thamama)	Upper (Asmari)	Lower (Bangestan)
Specific gravity at 15·6 °C (60 °F)	0·864	0·822	0·860	0·909
API gravity	32·4	40·8	33·1	24·2
Sulphur content (% wt)	1·79	0·92	1·50	3·02
Asphaltenes (% wt)	1·9	0·07	0·4	7·0
Pour point (°F)	20	0	15	−21
Wax content (% wt)	6·9	7·8	6·9	4·7
Vanadium content (ppm)			25	74
Nickel content (ppm)			8	19
Distillation products				
Gasoline (% wt)	15·6	22·1	15·9	10·3
Kerosene (% wt)	14·0	17·1	13·0	12·3
Gas oil (% wt)	19·3	21·6	19·4	18·0
Residue (% wt)	50·7	38·3	51·0	58·9

TABLE 3.3
Analyses of natural gases produced from various commercial fields
(% volume of components)

	Field					
	Groningen (Holland)	West Sole (North Sea)	Lacq (France)	Hassi R'Mel (Algeria)	Zelten* (Libya)	Sui (Pakistan)
Hydrocarbons						
Methane, CH_4	81·9	94·1	69·6	79·5	64·5	90·1
Ethane, C_2H_6	2·7	3·1	3·1	7·5	21·0	0·85
Propane, C_3H_8	0·38	0·58	1·1	2·5	8·4	
Butanes, C_4H_{10}	0·13	0·19	0·6	} 5·0	4·2	} 0·46
Pentanes and Higher hydrocarbons $|C_{5+}$	0·08	0·20	0·7		1·9	
Non-hydrocarbons						
Hydrogen sulphide, H_2S			15·3			
Carbon dioxide, CO_2	0·8	0·53	9·6	} 5·5		4·5
Nitrogen, N	14·0	1·3				3·5

*Feedstock for Brega liquefaction plant from oil production facilities

gradient. In abnormal cases the gradient can be nearly double or only half the average. Study of temperatures is important, for the generation of oil and its degradation to gaseous hydrocarbons and solid residues are related to temperature. Empirical evidence from hundreds of thousands of wells throughout the world shows that an oil domain normally passes to a gas domain somewhere between 15 000 and 20 000 ft. Knowledge of local temperature gradients is important for predicting the critical limits for the likely existence of oil or gas in any one area.

The pressure regime in the subsurface system is closely related to the hydrostatic head of the fluids in the pore space up to the surface of the earth. These fluids will normally be fresh water in the upper few hundred feet and saline water, sometimes saturated brine, at depth. The increase of pressure with depth is about 0·5 lb in² for every foot of depth. Thus at 10 000 ft the pressure in the pore space is about 5000 lb in². Deviations from the norm are caused by a variety of effects. Unconsolidated sediments can, by their own weight as overburden, apply a pressure which may more than double the hydrostatic pressure. Artesian effects from a distant, elevated aquifer, i.e. a continuous aquifer cropping out in a mountain range, can affect pressures below the surface at horizontal distances of scores of miles. Another cause of abnormally high pressures is related to tectonic pressures in areas of recent or present-day movements. Such pressures are sometimes transmitted to the pore fluids and retained. Very large accumulations of oil, and more particularly of gas, can cause pressures to be higher within the accumulation than in the hydrostatic regime surrounding them. This effect is due to the lower specific gravity of the hydrocarbons in what is best considered as a U-tube closed on one side by the sealing cap-rock and subjected to pressure on the other leg by the hydrostatic column of water. The 'oil leg' of the closed, or sealed, side of the system contains fluid lighter than the water; hence the pressure gradient in the hydrocarbon leg differs from that in the water-filled or 'open leg' of the system.

Knowledge of the pressure regime is important in understanding the movement of underground fluids. The location of some oilfields is determined by the hydrodynamic gradients and can be predicted from a knowledge of them. Knowledge of pressure is important in the development of oilfields and is of great importance in the drilling of wells. Correct assessment of pressures is vital in planning the casing requirements of oil and gas wells, the safety equipment on wellheads, and the density of drilling muds to be used to control the pressures at depths.

3. The role of stratigraphy

Having considered the fluid regime we turn now to the solid rocks. Assessments of the lithologies of reservoir rocks in all the world's major oilfields show that 58 per cent occur in sandstone reservoirs and 42 per cent in carbonates. A similar analysis of the world's major gas fields shows 75 per cent in sandstones and 25 per cent in carbonates. The Middle East has a profound effect on all statistics relating to oil. The existence of enormous carbonate reservoirs in the Middle East largely determines the overall balance by volume of oil in limestones and sandstones, but by numerical count there are in the world many more fields, both oil and gas, in sandstones than there are in carbonates.

Turning again to porosity and permeability, it is obvious that knowledge of, and the ability to predict, their values and their variations in space and time is most important, both in oil exploration and also in assessing and developing known fields. Predicting such variation cannot be done reliably by simple extrapolation or statistical methods and calls for an understanding of the geological processes that give rise to the changes.

The assessment of oil potential in new basins or the unexplored parts of existing producing basins calls for understanding of the full range of stratigraphic history and processes. Oil is produced from rocks of all ages from the Cambrian onwards. Small quantities of oil and gas are produced from metamorphic and igneous rocks of all ages, including the Pre-Cambrian. Although this production is relatively unimportant, study of the 'basement' is often part of the oil geologists' job, particularly in relation to the original supply of suitable clastic rocks to form reservoirs in the overlying sediments.

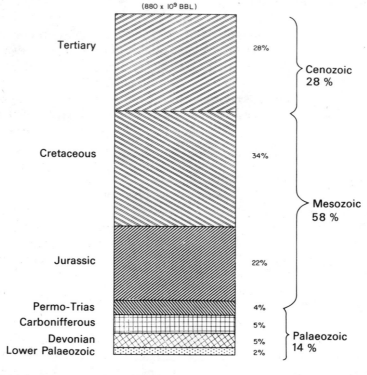

Fig. 3.2. Percentage of world estimated recoverable oil found in reservoir rocks of different ages (to the end of 1973).

Figs. 3.2 and 3.3 show the age distribution of reservoirs of all known oil reserves and their regional distribution, demonstrating that Jurassic and younger rocks contain some 84 per cent of the world's oil. It is worth considering the reasons for this and its implications. Accumulations of oil of commercial size depend on two factors other than porosity and permeability. One factor is the chain starting with organic source material and the diagenesis resulting from time, depth of burial, temperature, and so forth. There is clear evidence that the conditions necessary for this sequence have existed throughout Phanerozoic time. Within the context of this long time-span, only short intervals are required for the development of oilfields. It thus seems that this aspect of the development of oilfields is not particularly sensitive to time.

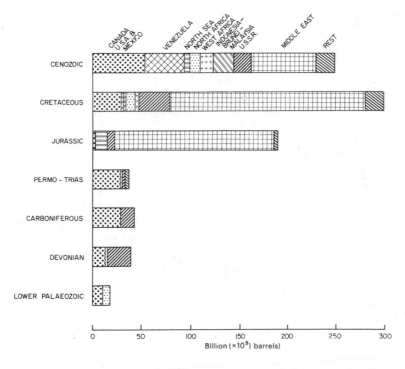

Fig. 3.3. Regional breakdown of estimated recoverable reserves of oil found in reservoir rocks of different ages (to the end of 1973).

Oil and natural gas

Another important chain of events is the migration of oil and gas into traps and the subsequent preservation of oil within these traps. Some degree of deformation is required to form the anticlines, unconformities, and faults that form the traps in which the oil is commonly held (Fig. 3.4). Too much tectonic disturbance with excessive tilting and faulting after the formation of oilfields results in loss of oil to the earth's surface.

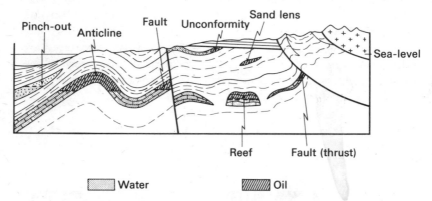

Fig. 3.4. Schematic diagram of typical oil traps. (By permission of the Royal Geographical Society.)

The optimum tectonic environment for oil is therefore one with a single phase of moderate deformation after a period of continuous shallow-water marine sedimentation. In many parts of the world most sediments older than Mesozoic have been subjected to several phases of tectonism or diastrophism, in many instances accompanied by metamorphism, and much oil, which had previously accumulated, has been lost. The main reserves of Palaeozoic oil occur in the U.S.A. and Russia in large sedimentary areas that have been stable continental masses since late Palaeozoic time.

Another obvious reason for the predominance of Jurassic and younger oil is that the current phase of ocean spreading started in the late Jurassic after the break-up of the supercontinent (Pangea). This has led to the continued existence of active zones of shallow-water sedimentation around the continents and the deformation of many of these piles of sediments to form large traps. Even within this phase the severe tectonism of some belts of sediments (e.g. the Tethyan (Alpine–Himalayan) and American Cordilleran systems) has led to the loss of much oil. One of the reasons for the high yield of oil from many shallow-water seas on the continental shelf is the existence within them of downwarped and downfaulted piles of gently deformed shallow-water sediments.

From the foregoing it can be seen that oil geology is concerned with rocks covering the whole of Phanerozoic time but with a preponderance of interest in the Mesozoic and Tertiary systems.

Considerations of age bring us to the role of palaeontology, which has an important role to play. In the drilling of wells and oilfield operations palaeontology has a major role in the dating of sediments drilled and the correlation of sections encountered to provide the basis for structure analysis. For the work on borehole material the obvious size limitations lead to a primary emphasis on micropalaeontology and palynology. Another major role of palaeontology is in the essential studies of sediment facies that lead to the understanding and prediction of reservoir qualities. Knowledge of the ecology of faunas, particularly of specialized facies-related faunas, is widely used in sediment studies within the oil industry; and for palaeontologists it provides a welcome diversion from routine identification and age-determination.

4. The role of structure

We considered earlier the hydrocarbons generated in the fluid regime of the pore spaces of sedimentary rocks and should now give some thought to the accumulation of oil into the segregations which constitute commercial oilfields.

Nearly all hydrocarbons of commercial importance are lighter than water and in an aqueous environment will float upwards. Significant rates of relative upward movement can take place only in strata with adequate permeability, and the path of upward migration is therefore usually up-dip rather than vertically.

It is obvious that unless the upward migration is arrested oil and gas will reach the surface. In the course of geological time vast quantities of oil must consequently, by dispersion by winds and waves, by oxidation, bacterial degradation, etc., have been removed from the inventory of fluid hydrocarbons.

The formation of oil and gas fields is largely a matter of arresting upward migration. The barriers to upward movement can either be stratigraphic changes that terminate a permeable bed or structures that juxtapose permeable and impermeable beds in trapping configurations. As the barriers to permeability are required in more than one dimension to produce a significant vertical development of oil, some element of structural deformation is required for the formation of most traps, except where a lenticular permeable body exists (Fig. 3.4).

The emplacement of oil in the traps as illustrated is probably not as simply explained by gravity differentiation as is often claimed. A good deal of movement of hydrocarbons probably takes place with moving water. An initial movement of fluids is obviously associated with the expulsion of pore liquids on compaction of the argillaceous rocks. Such rocks in an unconsolidated state have a very high fluid content and the normal compaction of these incompetent rocks will cause a flow into the more competent and permeable aquifers. Active flow over hundreds of kilometres also takes place owing to the charging of aquifers with meteoric waters in elevated topographic positions. This classic artesian flow is demonstrated in many extensively drilled basins where the pressure gradients and flow paths are accurately documented.

Several workers invoke the flow of water as the main source of transport of oil into traps. For this mechanism to work, the sealing or cap-rocks must allow preferential passage of water and the retention of oil. There is some supporting evidence for this mechanism but it is not possible to go into the detailed reasoning and evidence here. Nor is it possible to go into other complex and detailed arguments for the transport of oil in various chemical states.

The illustration of trap types as shown in Fig. 3.4 emphasizes the need for understanding structure in the search for oil, in determining the sizes of oilfields once they are discovered, and in locating production wells. In the finding and early phases of development of oilfields imaginative structural analysis is of paramount importance. Many important oilfields were found because of conjectural conceptions based on fragmentary evidence, both stratigraphic and structural.

Modern technology makes it possible to drill wells to some 34 000 ft. Accurate structural control is thus required to great depth. This naturally brings us to the role of geophysics, an all-important tool for oil explorers and producers. In the early days of exploration on land, much drilling was carried out on anticlines and other structures determined by surface mapping. Subsurface complications, even in large surface structures, usually require geophysical control of such features as the hade of axial planes of folds, the trace of thrust-faults, the effects of unconformities, and variations of stratigraphic thickness, all of which affect the location of deep traps. Geophysics can clearly be the only structural guide before drilling in areas where the deeper geology is hidden by unconformities, by superficial drift, or (especially) by water. Although various geophysical techniques are used for determining structure, including magnetic, gravity, and telluric (study of natural or induced electric currents), by far the most important tool is seismic survey. Of the two main seismic methods, reflection seismology is predominantly the main system employed. It is not possible to do justice here to the complex and detailed systems used for obtaining, processing, and interpreting seismic results but their paramount importance to the oil geologist must be stressed.

Understanding of detailed structure plays an important role in the development and drilling-up of oilfields. A production platform in the northern North Sea normally caters for 25 to 50 wells. Such a platform and all its facilities costs several hundred million pounds and if it is wrongly located and is consequently unable to reach parts of the oilfield owing to unseen faults or unexpected changes of dip, the economics of the venture can be very adversely affected.

5. Geological activities and techniques in the oil industry

Having attempted to give a feeling for the nature of oil occurrence and the relationship of some of the facets of a geologist's knowledge to the oil industry's concerns, let us turn to the more specific role of geologists in the industry.

Practices differ between various companies and sectors of the petroleum industry but there is a tendency to separate personnel working on exploration from those working on appraisal and production operations. Part of this separation is necessary but it would be quite wrong to think in terms of two virtually separate career structures.

The operations of exploration geologists have two distinct phases and the two types of people carrying them out tend to be in different stages of their careers. An exploration venture typically begins in somebody's mind, and it will either stem from, or initiate, a detailed geological appraisal based in the first instance on study of the literature. This study will normally call for a comprehensive

Oil and natural gas

review of all the known geology of a basin and will require draughting and map-making services to be available as well as publications and maps. Because of this, and because the decisions to be made will need to be taken by senior management, this activity is normally carried out either in a company headquarters in a capital city or in a main operating centre. Many of the people involved in this work are senior staff with at least ten to fifteen years' experience. More junior staff with only a few years' experience are also employed in the compilation work for such large-scale studies. The studies of basin prospects utilize all the methods available, including field survey, acquisition of data from foreign centres, photogeology, palaeontology and, where appropriate (and it nearly always is), the whole range of geophysical methods.

Although field survey work is becoming less appropriate in many areas of operations, particularly in the offshore areas, it is still important and widely carried out by companies. The British Petroleum Company has, for example, in recent years had geological parties out in many countries, including Alaska, New Guinea, Indonesia, Iran, Peru, and Colombia. There is no doubt whatever that an adequate understanding of geology and its significance world-wide can be obtained only by seeing a lot of geology in a lot of places.

The work of geologists in operating and producing areas is often thought of only in terms of subsurface geology or the rather derogatory term 'well-sitting'. Certainly the detailed examination of cuttings and cores from wells is an important aspect of development work. Geologists are present on all drilling operations and by carefully following the stratigraphic and structural progress of the well they plan its progress and completion. As offshore wells now cost up to several million pounds each, their safe and effective guidance to pre-planned objectives is important. The control of wells also commonly requires the use of various correlative methods including palaeontology, analysis of electric and radiometric surveys, and detailed lithological logging.

In addition to the control of drilling operations geologists in producing areas are engaged in many other tasks. They are concerned with the selection of well locations (an activity normally closely connected with geophysics and geophysicists), with the assessment of drilling results, the determination of structure, and with the assessment of the size and potential of oilfields discovered.

In reservoir assessment specialist petrologists are often employed.

Other activities in petroleum operations that call for geological expertise include the study of the uppermost sea-floor sediments in connection with pipeline routes, assessment of bearing capacities for producing platforms, etc. Geologists in oil companies also become involved in problems related to the shallower geology onshore. Apart from the obvious pipeline and foundation problems another challenge is in connection with underground storage systems. Underground storage is in two main types of cavities. The shallower systems are excavated galleries in hard rocks, usually in the uppermost two hundred feet of solid rock and numerous problems are encountered, including variations of permeability, distribution of fractures, and movement of groundwater. The second type of cavity is formed by underground solution of salt and this calls for accurate definition of the salt bodies involved.

6. Geologists in other roles

There are other vocations for which geological training is an obvious advantage.

Geophysics is a natural opportunity for geologists. Until recently there were few courses specializing in geophysics and most recruits into the industry came from the disciplines of physics and geology; both skills are required by most geophysicists. Those physicists who are engaged in instrumental work and on the computer processing of seismic records need a fuller basis of physics and mathematics. Many, if not the majority of geophysicists, are involved in interpretational work that requires a good background of geology and a lesser breadth of physics.

Petroleum engineers or reservoir engineers also need a background of geology, and geologists with some inclination towards physics and mathematics have considerable opportunities in these fields.

7. The future for oil and the oil industry

7.1. World-wide

There has been growing awareness over the past few years of the problems of consumption of non-renewable resources and the exponential patterns of demand of 'growth economics'. It is consequently natural to question the future of the oil industry. I have myself for some years been prophesying problems ahead on oil supply and price in the face of continued increases in demand at the rates of the recent past, and in the light of finding rates and apparent world availability for future discovery. But oil supplies are not in imminent danger of drying up nor are the oil and gas industries about to go out of business. Our mechanical and mobile societies are too firmly committed to oil and its use to do without it for many decades, and other energy supplies are beset with problems of cost, both economic cost due to dwindling reserves and inflation of labour costs, and social cost in terms of danger, environmental damage, and so forth. Large changes in the use of energy cannot be made quickly in response to the denial of one particular source of energy for reasons of cost, politics, or war. The small actual response to the fivefold increase in oil prices since October 1973 demonstrates this inflexibility of our use of energy.

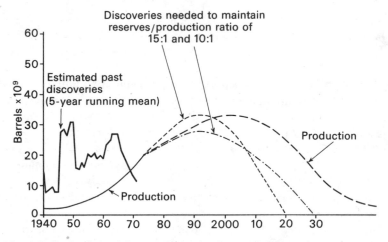

Fig. 3.5. Projection of future oil production and discovery requirements, assuming ultimate discovery in the world of 2000×10^9 barrels and production increasing by 2·25 per cent per annum until 1985.

Fig. 3.5 shows a prediction of the likely future pattern of oil production based on one set of estimates of total reserve availability and likely future finding rates. Although estimates of ultimate world reserves cannot be made with any precision and future finding rates are equally uncertain because of economic and political factors, the general form of the total production curve is reasonably predictable. The present stock of reserves in the world, coupled to our present dependence on oil, make the first five years or so reasonably certain. After that combinations of lack of available oil, price relative to alternative energy sources, political restrictions, etc., may cause the curve to flatten earlier. Some consider that lack of available reserves alone will cause a peak of production to be reached not later than 1985. Even under the most pessimistic assumptions there lie ahead many decades of oil production. Oil will still be valuable and worth finding well into the next century.

The finding of the reserves for the tail-end of the production curve will require more diligent search for the increasingly less obvious accumulations, and there will be no shortage of work for explorers nor for the geologists and reservoir engineers involved in the many secondary recovery operations that will be necessary to get the most out of declining reservoirs.

7.2. Oil in Britain

After being a country with relatively insignificant production by world standards, the United Kingdom is now entering the ranks of the major oil- and gas-producing countries.

Exploration for oil onshore in Britain started just after the First World War, but no significant production was found in a limited programme lasting sporadically for a few years. The vesting of all rights to subsurface hydrocarbons in the Crown and the introduction of tax concessions for domestic crude sparked a renewed, albeit modest, exploration programme in the 1930s that led to the finding of commercial oilfields in the Midlands in 1939. Very modest production and intermittent exploration has continued onshore ever since from a handful of fields in the Midlands and on the coast of Dorset. The leader in all these phases of exploration and production in the U.K. was BP (and its domestic predecessor D'Arcy Exploration).

The international industry was not attracted to the British scene until (a) the discovery of gas on a large scale at Groningen in Holland, (b) the verification of the Geneva Convention and the agreement on median lines by the countries bordering the North Sea, (c) the availability of drilling and production equipment for exposed offshore areas, and (d) the issuing of licences. This combination led at the end of the 1950s to an active programme of gas exploration in the southern North Sea. In the early 1960s this programme led to the discovery in British waters of a succession of

Oil and natural gas

major gas fields that now supply almost all of Britain's gas needs. Although most, if not all, of the major fields of the southern North Sea gas province have probably been discovered there certainly remains some gas to be discovered in small fields. Gas exploration drilling has virtually ceased as a result of the British policy on gas pricing, but this must change and renewed exploration will then follow.

As activity spread northward in the North Sea oil was found in large quantities, initially in Norwegian waters, in the separate northern North Sea basin which underlies deeper and more exposed waters. This led to a rapidly expanding and large-scale effort by the oil industry that revealed a series of most important oilfields. The total oil reserves found in the North Sea up to 1974 consist of some 18 thousand million barrels of recoverable oil, of which some 14 thousand million are in the British sector.

It is expected that the likely levels of oil production in the British sector of the North Sea will be on the following lines:

	1980	1985	1990	2000
Millions of barrels per day	2.01	3.0	2.0	1.5

To put these figures into perspective, Table 3.4 shows the major oil-producing areas in 1973. It can be seen that Britain will indeed be among the ranks of major oil producers and will continue to be so for a long time.

TABLE 3.4
Oil production from major oil-producing areas

	Production (Millions of barrels per day)
U.S.S.R.	9·2
U.S.A.	8·9
Saudi Arabia	8·4
Iran	6·1
Venezuela	3·0
Kuwait	2·6
Nigeria	2·3
Iraq	1·8
Abu Dhabi	1·8
Libya	1·7
Rest	10·9
Total	56·7

It is difficult to quantify the long-term profile of British oil and gas production and activity, for it will depend a great deal on government policies and politics. Given the right political and economic climate there is little doubt that the British involvement could, from the basis of the size of the inevitable short-term development and activity that is under way, become one of the world's leaders in petroleum exploration and production technology. The world-wide dominance of the United States oil industry resulting from the 'export' of its domestic experience is an object lesson.

Whatever the future holds for the British oil and gas industry there is little doubt on a level of demand for skilled personnel far in excess of past opportunities. The total number of oil geologists, geophysicists, and petroleum engineers working in the U.K. was about three or four hundred twenty years ago; now it is three or four thousand. Although the balance will change with a large proportionate increase of reservoir and production engineers the numbers employed by the industry will increase rather than decrease. The setting up of the British National Oil Corporation will entail the recruitment of several hundred oil technologists of various kinds.

Acknowledgements

I should like to acknowledge the help received from various of my colleagues in British Petroleum and, in particular, Miss Anne Sayer who prepared Figs. 3.2, 3.3, and 3.5, and Dr. G. Spears, who provided the information for Tables 3.1, 3.2, and 3.3.

4. Coal

GEOLOGY APPLIED TO SUBSURFACE MINING

1. Historical

The evolution of geology and coal-mining technology have been interrelated from the seventeenth century onwards. Early workers included George Sinclair who, in 1672, proved the essentially synclinal nature of the Mid-Lothian coalfield, Robert Plot who, in 1687, outlined the geometry of the North Staffordshire coalfield, and John Strachey who noted in 1719 the disposition of the major seams of Somerset and moreover interpreted the concealed part of that coalfield (Bailey 1952, pp. 1–4). In 1790 William Smith was acting as consultant to the Tarbock Colliery in Lancashire and in 1819 he advised on two mining ventures for Jurassic coal in Yorkshire. During the early nineteenth century coal-mining expanded enormously and such expansion demanded accurate geological information. That demand was a great stimulus to the development of the science and it is significant that in the U.K. the Geological Survey (now part of the Institute of Geological Sciences) and a number of regionally based geological societies within the coalfields were formed at that time. It is also not surprising that a number of coal owners and mining engineers figured prominently among the co-founders of, and early contributors to, the geological societies. All this time, however, and indeed until recently, the primary tasks of the geologist were concentrated on the classical aspects of field mapping, correlation, and elucidation of coalfield structure.

On the nationalization of the greater part of the coal industry in 1947 the then slender geological resources of the National Coal Board were largely occupied in proving new reserves to balance the approaching exhaustion of many of the existing collieries. This initial work was primarily concerned with exploration geology; it entailed drilling deep boreholes and siting of several new production units. At the same time, the Coal Survey, a branch of the

Scientific Department of the National Coal Board, inherited the task of the former Fuel Research Board in carrying out a systematic chemical survey of the British coal resources. Their work, embracing coal geology as distinct from coalfield geology, has contributed to the evaluation of seam qualities for determining their most applicable technological usage. While such work is invaluable to the industry, and moreover has resulted in an accumulation of critical data on coalification processes (Teichmüllers 1968; Williamson 1967, pp. 221–6), it is outside the scope of the mining geologist's work as defined here.

By 1960 the emphasis from 'coal at any price', as determined by post-war exingencies, was changed, in the light of what are now thought by many to have been a series of mistaken governmental policies relying on foreign fuel sources. In 1965 considerable restrictions were placed on the industry: the sinking of new collieries to replace the production of old or exhausted pits was decelerated; many uneconomic collieries were closed and exploratory work was curtailed.

The mining industry itself to some extent countered the effects of colliery closures and a shrinking labour force by an intensive mechanization programme (Table 4.1). Accordingly the industry has changed from labour-intensive to capital-intensive operations. Over 90 per cent of coal production is now from mechanized cutting and loading equipment on coal faces with self-advancing roof-support systems. Underground tunnels are similarly becoming increasingly advanced by mechanized heading machines. The use of such machinery, the cost of which on an individual coal face may exceed £800 000, is economically dependent upon uninterrupted production. The emphasis has therefore changed to exploitation geology concerned with predicting changing face conditions and assessing the viability of mechanized mining in undeveloped areas of collieries. The sophisticated technology of modern mining is much less flexible than the former hand-worked systems, and reserves within a colliery 'take' that were formerly economically viable may now be considered unworkable. Thus a fault of relatively small displacement, which would be merely a nuisance on a hand-worked face supported by a simple 'prop' system, might result in the total abandonment of a highly mechanized coal-face. Since the trend is towards a decrease in number, and a corresponding increase in the overall size, of production units daily face yields of 2000 tons could be normal (Browne 1967, pp. 825–30). The failure of such a large unit cannot always be offset against the remaining productivity of the other three or four faces within a colliery. In that event the colliery itself may close and several thousand men become out of work.

TABLE 4.1

Trends in productivity, manpower, and mechanized output, 1947–72

Year	Number of producing collieries	Deep mined output (million tons)	Output per man-shift (cwt.)		Mechanized output (%)
			Overall	Face workers	
1947	958	184·4	21·5	58·4	2·4
1952	880	210·6	24·2	64·2	4·9
1957	822	207·4	24·9	67·6	23·0
1962	616	187·6	31·2	91·0	58·8
1967	438	164·6	36·6	113·7	85·7
1972	289	109·2*	41·9	138·8	92·2

*Estimated 26·3 million tons lost by industrial dispute
(From N.C.B. Reports and Accounts 1971–2, Vol. 2, Table 11. National Coal Board, London)

2. Geological constraints in exploitation

The major geological constraints to viable mining are related to the sedimentary frameworks and tectonic setting of the British coalfields (Trueman 1954; Clarke 1966). The principal sedimentary constraints are reflections of the varying natures of both the interseam strata characterized by strong cyclothemic developments and the coal seams themselves (Duff, Hallam, and Walton 1967).

A primary parameter in the determination of the mining layout and choice of equipment is that of seam thickness which, apart from such notable exceptions as the Warwickshire Thick Coal formed from the local coalescence of a number of individual seams, seldom exceeds 3 m and usually varies between 1 and 2 m. A general correlation exists between seam thickness and colliery output (Armstrong and Clarke 1970, pp. 422–3) although the thickest seams do not necessarily yield the highest outputs. Seams exceeding 2 m present difficulties with modern powered face-support systems and those less than 0·75 m are not ideally suited to mechanization. Thickness variations may result from the development of seam splitting, often towards the centre of the depositional basin, or by facies changes in the roof strata. Again, while the general seam thickness may remain constant, the coal type may alter as a result of localized variations in the original depositional environment. Although the replacement of 0·25 m of 'top coal' by a similar thickness of cannel, formed within a contemporaneous lake, may have little effect upon the face output the technological usage of the product may be adversely affected.

Most, if not all, British coals are affected by seam split phenomena (Broadhurst *et al.* 1968; Raistrick and Marshall 1939) largely caused by differential subsidence within the coal-forming basin and to a lesser and more local extent by channel migration of contemporaneous rivers. The rapid onset of splitting, usually heralded by a progressive increase in thickness of non-carbonaceous or 'dirt' partings, may result in particularly bad roof conditions above the lower member of the split. Furthermore, even if several members of the split attain workable thicknesses their close proximities may permit the successful extraction of only a single unit. The technological properties of the coals within the split may be dissimilar from those of the entire seam. At Burnley, in the Lancashire coalfield, the Union Seam is formed by the coalescence of the Lower Mountain and Upper Foot Seams. The former is characterized by very low ash and sulphur contents and yields an excellent metallurgical coke. The Union Seam, however, exhibits the high sulphur content elsewhere associated with the Upper Foot Coal and is therefore suitable only for use as fuel.

In addition to adversely affecting the quality of a seam the 'dirt' may have correlative applications or affect safe working. The rare kaolinitic tonsteins that usually occur as thin partings within the coal and are generally accepted as resulting from contemporaneous volcanicity form some of the most reliable correlative horizons (Williamson 1970). Quartzlagen bands composed of cryptocrystalline quartz precipitated within the seam (Hoehne 1956) may result in the generation of 'hot' sparks when cut by machinery and so cause ignition of methane. The risk of such incendive sparking is highest in rocks containing over 30 per cent quartz and with grain sizes exceeding 5 μm (Pearson 1974, p. 323). Sparking may also occur in the machine mining of coal containing pyritic inclusions. Incendive sparking related to the occurrence of limestone con-

Coal

cretions containing pyrite masses within the Union Seam at Burnley resulted in the recent closure of a colliery.

The immediate roof and floor conditions exert a considerable influence upon the selection of equipment, the design of the face layout, and the efficiency of the coal-face (Clarke 1966, p. 34; Mills 1972, p. 236). Numerous examples from the East Midlands coalfield are given by Elliott (1974, Table 1) of adverse conditions created by sedimentary structures upon the coal-face.

Variations in the natures of the seat-earths ('fireclays' and ganisters) forming the floor strata present severe problems in the maintenance of supply and haulage roads leading to the coal-face and in the operation of the support systems upon the face (Pearson and Wade 1967). Many roadways require almost continuous programmes of remedial treatment to maintain their size and so permit adequate face ventilation and the movement of men, supplies, and coal. Such roadways are commonly affected by 'heaving' of the basal underclays as a consequence of the high pressures, resulting from coal extraction, induced upon the protective packs at their sides (Whittaker 1974). Should the upper sections of the seat-earth be composed of relatively strong material overlying a softer unit, and separated from it by a slickensided zone (Wilson 1965a, 1965b), tensile failure results in the tilting of planar masses of the floor.

Alternatively, and particularly in the presence of water, a soft seatearth may fail by pseudo-plastic flow. The latter, close to the coal-face, may cause tilting and floor penetration of the roof supports and so result in a time-consuming and arduous extraction procedure. The effect may be accentuated by the occurrence of an overlying sandstone aquifer, rendered more permeable by induced fractures, and penetrated by 'roof breaks' from the face.

The natures and thicknesses of the various interseam strata within 10 m above the coal-face exert increasing effects upon the environment of the coal-face. Many roof-support problems resulting from excessive 'weighting' or roof pressures, which restrict normal advance of the systems, can be attributed to occurrences of weak partings at some distance above the seam. Such problems may occur when a comparatively thin coal 'rider', often resulting from seam splitting, is separated from the main coal by a few metres of seatearth or soft mudstone. Such parting surfaces are commonly developed as slickensided zones or highly listric thin clays and occasioned by tectonically induced bedding-surface slip.

In other cases potential partings may be situated at the bases of erosional sandstone channels. If the sandstones are water-bearing, troublesome and occasionally hazardous water feeders may result. Moreover such erosional 'rambles', confined to the roof strata, as distinct from the more easily identified washouts within the seam (Williamson 1967, pp. 193–8), result in abnormal stress conditions, causing local weighting of the face supports and buckling of the arches supporting the roadways (Clarke 1966, p. 34).

The nature and occurrence of many faults are usually well established at particular horizons within the existing coalfields. The mining techniques formerly used made more or less continuous extraction possible in moderately faulted ground, which therefore became well proved. However, lateral or vertical projections from such established fault positions towards new and as yet unworked horizons can lead to serious misconceptions. The variations of hade within sequences of differing competence may result in considerable displacements from the projected positions. Similarly many minor faults, of displacements less than a few metres, are often confined to the argillaceous members of the cyclothem and run-out in the basal parts of some of the thicker sandstones (Clarke 1963, p. 211). In other cases distinction should be made between localized compaction fractures, developed at the edges of sedimentary discontinuities and affecting strata control beneath such features, and true tectonic faults.

While a general correlation exists between the frequency of faulting and productivity, a less obvious and more important effect is determined by the size and orientation of fault-free blocks in relation to the width and direction of the coal workings. A coalface some 200 m long in an area of high regional fault intensity (Clarke 1966, pp. 36–7) but with a unimodal fault orientation parallel to the face advance may have a long life. In contrast, a similar face situated in an area of low regional fault intensity may be more likely to encounter faulting if the regional fault orientations are polymodal.

The natures and intensities of jointing in the surrounding strata may influence the location of supply roads leading to the coal-face, the width of the face, and the direction of working (N.C.B. 1972, p. 3). Obvious strata-control problems are occasioned by high joint frequencies in the roof strata and parallel to the tunnel direction or coal-face edge. Alternatively some of the thicker, sparsely jointed sandstones may form large unwieldy debris and so reduce tunnelling efficiency. A further constraint exerted by the tectonic setting, and of great importance to mine safety, appears to be a relationship between gas outburst fractures and a particular joint trend. Although coal-mining has always been affected by, and often induces, methane emission from both the coal and permeable interseam strata, the volume within the mine is normally controlled by ventilation and methane extraction boreholes. Occasionally, however, sudden and hazardous methane outbursts occur from the floor strata of particular seams. The floors in question are composed of a strong seat-earth or otherwise impermeable stratum overlying a gas-bearing sandstone or coal. Work by Arscott and Hackett (1969) suggests that such outburst fractures follow one of the regional joint directions. The safest conditions would appear to exist when the face advance and roadway heading directions are at 45° to the joint set.

3. Evaluation of the constraints

The day-to-day work of the mine geologist is concerned with the collection and analysis of geological data and the evaluation of the constraints upon mining. This work is essentially of a 'troubleshooting' nature and requires rapid appraisals in order to alleviate the effects of particular geological hazards. Thus, if a quartzlagen develops in the top coal of a face the mining engineer requires identification and an opinion on its likely continuity. Similarly an erosional channel, previously confined to the roof strata, may descend into the coal. A decision is required on the trend of the structure and its continuity within the coal. In certain cases it is possible, by shortening the length of the coal-face, to avoid the feature. Alternatively palaeocurrent analysis, by revealing a sedimentary trend at low angles to the coal-face, might indicate that the effect upon productivity might be relatively short-lived and so influence a decision to continue the advance of the entire face.

The geologist is becoming increasingly involved in long-term planning (Elliott 1974; Mills 1972). For this he is required to consider the entire 'take' of a colliery or coalfield area with regard to its future assets in unworked or only partly worked seams. Any previously encountered geological constraints are then reviewed and extrapolated into the virgin areas. Such considerations are subject to continuous reassessment as new data become available.

The classical geological interpretation of data accumulated from surface mapping, boreholes, and underground geological records is, by the nature of the science, in terms of probabilities. A communication problem consequently exists between the mining engineer, who is used to precise definitions and evaluations, and the geologist experienced in the infinite variety and gradations of phenomena and materials. The techniques of management decision are relevant here, and simulation exercises (Clarke 1967) are therefore practised by the geologist and the engineer, working in close co-ordination. The effects of several permutations of a series of possible geological constraints are considered, together with their financial implications, on a number of exploitation schemes. Such simulations will probably reveal the occurrence of areas within the proposed mining layouts in which the absence of definitive geological information will exert particularly critical effects upon exploitation. At this stage the positions of tactical boreholes or exploratory drivages may therefore be most advantageously located. While primary consideration may be given to one particular coal-seam the possibility of future developments within other unworked horizons would not be ignored. Thus it might be considered advantageous to continue drilling to some other seams and so add to the data bank for those horizons. Such considerations could outweigh the more immediate financial advantages of a series of relatively short underground boreholes as distinct from a surface drilling programme.

4. Assessment of reserves

It is important to distinguish those reserves estimated by the economist from those of the mining engineer. The latter would consider as reserves only those deposits which can be viably mined in the light of present technology, suitable geographic location, and current world prices. The economist, and to some extent the classical geologist, does not reason for the immediate present, or even the next decade, and so takes into account deposits whose usefulness is so remote in the future or those that are so geographically isolated that the mine operator would consider them entirely unexploitable. Various comments have been made about the 'immense coal reserves' of Antarctica, which are in terms of present economics, geopolitics, and conservation considerations, unexploitable. Similarly, although several seams may each attain workable thicknesses their close proximity within the vertical sequence may preclude the successful mining of more than one while favouring their profitable exploitation at very shallow depths by open-cast methods.

Current practice within the National Coal Board (Weekes 1972) is to exclude from reserve calculations those areas of coal in which a seam is (a) over 1200 m deep (the high bulk factor and current low price of coal as compared with other deeper mined minerals, normally excludes economic working at greater depths); (b) too tectonically disturbed to work; (c) less than 0·6 m thick; (d) too poor in quality, with ash contents exceeding 25 per cent; (e) affected by any future workings in an immediately adjacent seam which has already been included in the assessment; (f) too badly affected by vertically adjacent old workings; (g) too wet to work, by proximity to natural aquifers or flooded workings; (h) required to support important surface or underground structures. The overall reserves, calculated on volume/tonnage conversion factors of 1675 tons per foot per acre for anthracites and 1600 tons per foot per acre for bituminous coals, are further considered according to the degree of financial risk involved in their extraction. Accordingly three classes and a fourth, less reliable category are recognized (Weekes 1972):

Class 1. This lowest-risk category consists of these reserves with assured extraction rates, market values, and qualities. Over 90 per cent should generally be recoverable. There is a high degree of knowledge concerning seam thickness, depths, gradients, faulting, and the sedimentary environment and there should be no significant variations in the quality or output of the product. In general the reserves are so allocated only when the sedimentary framework has been proved from previous workings or drivages along two sides if the area is triangular in shape, or upon three sides if the area is a quadrilateral. In areas other than those with low regional fault intensities the reserves should be overlain or underlain by previous workings.

Class 2. The medium-risk category includes those coals for which the thickness, depth, gradient, quality, and geological environment are sufficiently well established to support the preparation of detailed extraction plans. Nevertheless the general level of geological uncertainty is too great to locate local hazards which would adversely affect the planned production or coal quality. No major shortfalls of production are, however, envisaged and in most cases over 75 per cent of the coal should be recoverable.

The sedimentary environment should have been established along at least one side of the area or proved by boreholes to include no major sedimentary constraint. The density of tectonic data is insufficient for the location or effect of local faults to be known with certainty.

Class 3. The reserves are associated with a high level of geological uncertainty so that exploitation is not plannable. Knowledge of the sedimentary and tectonic framework is insufficient to eliminate the possibility of washouts, seam splitting, and faulting resulting in the sudden and premature cessation of coal-faces.

Unclassified reserves. These include those currently considered non-viable as a result of marketing requirements and mining methods. Alternatively they may occur in areas with no underground provings although within a geological environment, as determined by surface mapping, suggestive of a future potential.

A classification of reserves is not a once-for-all-time exercise. Within the colliery it is essentially a summary of the assets or resources (Mills 1972, p. 238) which are *available* to a given production unit at a particular time. As such it should be periodically reassessed in the light of additional geological information derived from the mining programme and changes in the economic climate and marketing requirements.

5. Exploratory assessment

While exploitation geology is currently dominant the world energy crisis has already stimulated the exploration of virgin areas and the formation of at least one new coal exploration group. New collieries within the United Kingdom are being planned. The Selby coalfield, proved by exploratory drilling in the mid-1960s (Goosens 1973) and with estimated recoverable reserves of some 500 million tons, is now regarded as a prime development area (Mills 1975). Although exploration is chiefly concentrated within or upon the margins of the established coalfields, other areas, such as the totally concealed Oxfordshire coalfield (Poole 1967, pp. 17–24), are now receiving attention.

Such exploratory investigations, involving a more classical approach than that required for exploitation, require a progressive drilling programme. The exact sites of the initial boreholes are often determined by access or town and country planning requirements. Open-hole drilling may be practised in the near-surface cover rocks, but the lower beds of the concealing strata, which often contain important aquifers and the entire Coal Measures sequence, should be core-drilled. The introduction of the split inner-tube core barrel, plastic linings, and step-type diamond core bits have in the past decade enabled 100 per cent recoveries to be achieved. Additional information is provided from in-hole geophysical surveys. The gamma-ray, neutron, and formation density logs are in this respect particularly applicable (Laughton 1970, pp. 389–91) and enable accurate estimations to be made of seam thickness. At this stage some, albeit generalized, estimation of the coalfield structure may be established by correlative studies. The coal cores would be analysed and their technological properties determined.

A second, strategic phase of drilling would concentrate on proving seam continuity and the elucidation of any trends of seam splitting and coal impoverishment. The phase may be supported by surface geophysical surveys as aids to the location of major faulting. Such land-based surveys have generally proved disappointing, for the repetitive cyclothemic developments possess insufficient geophysical contrasts over wide vertical ranges. The methods are, however, applicable to the location of faults with major post-Carboniferous components in the concealed coalfields. On the completion of the strategic phase and with the experience of simulation exercises a third phase of productive drilling would be required to prove the structures of the apparently more viable areas.

The core-drilling equipment used is of the same basic design as that employed for engineering site investigation but coal exploration requires greater drilling capabilities, sometimes exceeding 1100 m. At such depths normal core drilling is slow because of the repeated pulling up and lowering of the drill-string after the core barrel, usually 6 m long, has been filled. The development of the wire-line equipment has considerably increased drilling rates by eliminating the need to remove the drill-string. The equipment consists of special drill rods having outer diameters close to that of the borehole. Within the assembly is an inner core barrel which can be independently lifted to the surface. The technique is becoming increasingly used in the British coalfields and has already been utilized to a depth of 2400 m on the Continent. Similarly considerable use is now being made of 20-m core barrels in deep boreholes; in this way the number of 'round trips' is reduced enabling more rapid borehole completion.

6. The useful geologist

Coalfield geology requires academic versatility, a strong practical interest, an awareness and appreciation of mining and economic requirements, and, above all, an ability to communicate on inter-disciplinary levels. While an academic awareness of the subject is important it should not become detached from the realities of economic existence. For the coalfield geologist a fault is not merely a fascinating example, it is also a potential restraint upon successful coal production.

The coalfield geologist should have particular expertise in the sedimentary and structural aspects of his science, a knowledge of correlative methods, ability in field mapping, and preferably a working experience of geophysics and an understanding of basic hydrogeological principles. Furthermore, in areas of shallow mining he requires an understanding of the natures and types of superficial deposits and geomorphological development. Acquaintance with rock mechanics is desirable and the ability to work, mix, and communicate with men of all types is essential.

GEOLOGY APPLIED IN OPENCAST WORKING

7. Historical

Most historical books on coal-mining quote examples of the open-cast working of shallow coal throughout the centuries, beginning with the Romans, but the tonnage extracted was very small. It is only since 1942, when opencast working was introduced as a wartime measure, that it can be considered as a significant industry in Britain. A profit was neither achieved nor considered necessary in those early years, though improved mining methods gradually closed the gap. At about the time that the coal shortage ended, opencast working started to show a profit. In general, this was achieved by more efficient methods, the outstanding one being the use of bigger draglines for the excavation of overburden. The opencast organization, which had previously come under the Ministry of Fuel and Power, was in 1952 transferred to the National Coal Board.

Although opencast working proved profitable, there was much opposition to the use of this method of extraction. It was argued—and disproved—that opencast coal was inferior in quality; re-storation was criticized but this ignored the vast strides taken since the war whereby restored agricultural land was as good as or better than before; and the comparative safety of opencast working received little publicity. One detrimental factor that obviously could not be denied was the temporary disturbance of the land. In 1959 it was decided by the Government that, largely to protect employment in the deep mines, opencast work would cease apart from exceptional sites, which were mainly in South Wales, for working anthracite, of which there was still a continuing shortage.

After a few years, there was a gradual increase in opencast to be succeeded by another setback in 1967, when the Government restricted output to special coals which were not in competition with deep mines. By now, however, this category included coking coal as well as anthracite and so the situation was not as severe as in 1959. Also the Opencast Executive was pressing forward with the idea of clearing up derelict areas in association with coal-winning so that the organization which had been accused of dereliction was able to show that it could, in fact, improve the surface of the land. Profitability improved during the 1960s and production moved upwards.

With the sudden occurrence in 1974 of not only a national shortage but also a world shortage of coal, the proposed target for the Opencast Executive moved to a higher level: of 15m tons per annum.

Reserves of comparatively shallow coal are now being sought in nearly all the exposed coalfields of Great Britain. The most important economic factor in opencast working is the average ratio of overburden to coal. Broadly speaking, at 15:1 (or somewhat higher for anthracite) such reserves can be worked at a handsome profit. At 15:1, it is not unusual to dig coal from depths of about 130 m; in one exceptional case in Scotland digging is going down to over 200 m. Apart from the ratio there are, of course, other difficulties in excavating to great depths and although it is possible to imagine sites within this ratio at depths of 500 m or more, it is not expected that these will be economically viable.

Although the general outline of the geology of Britain is well known, the detailed structure is virtually unknown and it is the details that particularly concern the opencast geologist. It is important to interpret minor faulting, folding, wash-outs, variations of coal thickness, and superficial deposits.

8. Investigation

Prospecting for opencast coal at comparatively shallow depths is relatively cheap. The average depth of working and, therefore, the average depth of drilling is only about 40 m. By spacing boreholes at average intervals of about 50 m the cost of proving coal is not excessive, being only 10p per ton of coal found.

One of the most important factors in estimating quantities in opencast work relates to abandoned shallow underground workings. These workings occur at about the depths where opencast mining is proposed. They are usually 'pillar-and-stall' workings and the pillars represent coal which can be recovered by opencast methods. There are normally no plans available for such ancient workings and at present the only method of determining the percentage of coal which has been extracted is by drilling. In practice it has been found that spacing boreholes about 30 to 50 m apart gives a reasonably accurate answer.

8.1. Development record maps

To select a potential site a geologist will need to study all available information. Basically this consists of geological maps, plans of underground and opencast workings, local knowledge, special field investigations, aerial photographs, and the results of previous drilling.

In the early days of opencast, this was a fairly simple matter but as the years went by potential sites became increasingly hard to find and it was necessary to approach the problem in a more scientific way. Some geologists are concerned with producing maps with the prime purpose of locating new sites worthy of further investigation. In Britain the work proceeds very much along the lines of the map work carried out by the Institute of Geological Sciences so that all available information is gathered together and interpreted. It should be noted that the most prolific source of British geological information is now in fact the Opencast Executive's own drilling: the Executive, and the Ministry before them, have been putting down some 15 000 boreholes a year for over 30 years. Where mapping is carried out by the I.G.S. geologist (rather infrequently these days) the opencast geologist will co-operate closely. Several maps have recently been produced jointly by the Opencast Executive and the I.G.S.

8.2. The geologist in the field

The geologist in the field relies upon the results of drilling for his accurate interpretation of structure. As drilling proceeds he, or she, will build up a geological picture and as this evolves in his mind (and on his plan) he will need to think not only in terms of structure but also of ratio (overburden to coal) and, therefore, in economic terms. Many other factors will have to be considered, such as the amount of hard rock, the severity of folding and faulting, the shape of the coal area (which itself can be an economic factor), the occurrence of major overhead and underground services, and the nearness of houses and other buildings.

The question of borehole spacing will influence him from the start. He may wish to have a 'skeleton' pattern which can be filled in later. It is, however, not always feasible to drill in a pattern for if the structure is very complicated or the variation of old working density and other factors are considerable, he may have to pursue the old adage of going 'from the known to the unknown'.

The geologist will need to consider the question of quality and to decide if a particular seam is acceptable; if it is not, the question then arises as to whether an underlying seam is within an acceptable overburden:coal ratio. All this combined with the over-whelming problem of old workings means that he has to decide carefully the position and depths of drilling of each borehole, the amount of coring, and also the cost of drilling. In practice, he will probably spend part of the working day on the prospecting site seeing the drillers and noting what they have found, measuring cores and deciding the location of future holes, returning to his office to consider problems of structure and keep plans up to date. After perhaps two or three months on an average site, controlling one or two drilling rigs, he will have drilled a coal area to its economic and geographical limits. He will then proceed to another prospecting site with completely different problems and completely different structure but with the same overall purpose: to put down the minimum number of holes to get the right interpretation of the geology while bearing in mind that the area drilled must be economically viable.

There are relatively few problems on working sites when extraction is in progress, but for his personal satisfaction the geologist will visit these sites and be able to see, often soon after the completion of his plans, whether or not the structure is as he had envisaged it.

8.3. Drilling

With drilling, as with draglines used for excavation of overburden, improved mechanization reduces costs, for the newer and larger drilling machine is less expensive per metre than the earlier smaller one.

The standard method of drilling in the early years (the war years) was the hand drill. The average depth of drilling in those days was less than 12 m and labour was cheap by comparison with today's costs, making the hand drill an ideal drilling tool for that period. A natural sequel to the hand rig was another percussive drill, the slip rope rig. This ancient method of drilling was very good—better than the hand drill because a larger chisel could be used. It had excellent mobility and would still be used today if it were not for the high cost of labour. Next was a power tool which was percussive and rotary: the water flush. In essence it was a jack hammer held in position either on a slide or on a rope. Finally, there was the true rotary rig, which is best described as being the traditional diamond drill. There was a gradual improvement in drilling performance over the years and in 1956 the air drill was introduced. The machine used had an hydraulic rotary drill head. The compressor as well as the drill were mounted on a lorry. The air drill gave excellent results and with such an improved performance that drilling prices were considerably reduced. It had the big disadvantages of being very heavy and of doing more than average damage to agricultural land.

The Opencast Executive therefore considered the possibilities of a lighter and more mobile drilling rig and finally introduced a tractor-mounted, top-drive rig which was developed by one of its drilling contractors. Other contractors followed suit, either with this machine or with developments of their own, and the Executive now mostly uses tractor-mounted drilling machines. The improved cross-country performance is very noticeable and it is possible to cross ground which previously might have been impracticable. One diversion from this general trend is the use on the Scottish moorlands of vehicles with wide rubber tracks whereby the ground pressure is reduced well below that of a tractor.

8.4. Geophysics

Drilling for opencast working requires comparatively shallow holes and the percentage of coring is small; the cost of drilling per metre is in consequence comparatively low. This must be borne in mind when considering the application of geophysics to opencast prospecting. Many geophysical methods are well known and well tried and some of them could (and have) been used for determining drift-filled buried channels, faults, unconformities, and other geological phenomena; nevertheless, there is still a need to drill.

A real benefit would come from a geophysical method which could accurately determine the percentage of old workings in a pillared area. This would result in real economic benefits, bearing in mind that the close pattern of drilling is mainly determined by a need to quantify old workings. So far, no geophysical method has been able to do this.

Electrical logging of boreholes is well tried and well proved. A density log (radioactive source) supported by a gamma-ray log will give a reasonably accurate interpretation of coal and other strata. Again the question of cost arises when comparing this method with the comparatively cheap drilling required for opencast work, but with improved techniques, logging may eventually become a more general method of prospecting.

8.5. Computation

There are two different approaches to computing and the question arises as to whether the benefits of using a computer will outweigh the additional cost as compared with traditional methods. The computation of quantities entails simple calculations (area × thickness = volume); the basic requirements are to determine the total excavation and volume of coal and other deposits. The traditional method of measuring areas is by planimeter and for total excavation by drawing isopachytes and measuring the areas between them. Various statistical analyses have to be done while considering such factors as coal thickness and quality, sandstone and seam interval thicknesses, and old workings. The computer can achieve much but some further benefit must be shown to justify the cost of using a computer. Such a benefit could be found in site evaluation. With any manual output, there is a certain amount of uncertainty as to whether the ideal area has been selected; and 'ideal' here means one which satisfies certain parameters. For instance, one might require the maximum tonnage at a minimum profit and this is where, in theory, the computer gains; it can quickly calculate alternative sites or combinations and select the one suited to requirements. Site evaluation is, however, a difficult problem; not only is it necessary to consider the geology and the quantities, but allowance must also be made for the method of working, the equipment used, and the effects of double handling at depth. If the geology were simple, the solution might be easy, but with complications of structure and methods of working the Site Evaluation System is a goal yet to be achieved.

9. Extraction

9.1. Stability of excavated slopes

All opencast sites have stability problems and, directly or indirectly, the geologist is involved. A minor slip within an opencast site is normally no problem; provided there is no danger to life and a minor slip does not develop into a major one affecting buildings or services outside the area, the only disadvantage is extra work, delay, and perhaps some loss of coal.

One important factor is time. Most opencast sites are short-lived and any particular face may be exposed only for a little while. Many failures are time-dependent and it is normally in large sites where exposures remain for a considerable period that real problems can arise. If the excavation is deep and there are several faces joined by haulage roads, then the influence on working can be considerable.

Every geologist needs to consider the elementary factors involved in slips. These can be summarized as follows:

Natural structure	Relationships between rock and drift
Geological structure	Bedding, jointing, and faulting
Materials	Nature of superficial deposits, weak rocks, clay mylonites, and fault gouge
Water	Surface, in drift, in strata, and in old workings
Boundary conditions	Laterally around the site, and beneath the site

Failures dependent upon the material forming the slope can be generally summarized as follows:

Fig. 4.1. Syncline exposed in opencast working, Westfield, Fife.

Rock material	*Soil material*
Plane failure—typically on undercut bedding-planes	Circular failure—typically in clayey drift
Wedge failure—typically on inclined joint sets	Creep—typically in soft drift at relatively high moisture content
Toppling failure—typically on steep joint sets	

The geologist in the field will be concerned with identifying the factors which could possibly bring about a failure; it is for the engineering geologist to carry out a detailed analysis. Where possible, conditions likely to give rise to slope stability problems should be identified at the exploration stage in order to assist in the programming of extraction.

9.2. *Geological conditions on sites*

There is a wealth of geological information available, not only from drilling but also from the excellent exposures which can often be seen as textbook examples on working sites and it is a matter of regret that so little use is made of all this geological information by anyone outside the opencast industry (Fig. 4.1). Only a few of the more interesting examples can be quoted here.

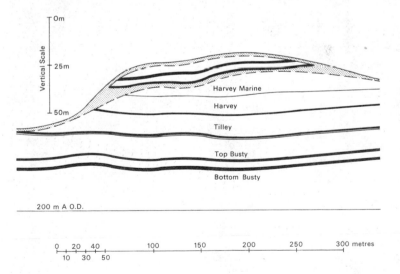

Fig. 4.2. Rafting of Coal Measures strata on hilltops near Tow Law, County Durham.

Coal

Near Tow Law in Co. Durham several prominent hill caps long thought to be formed from the sandstones overlying the Harvey seam were found to be rafts or erratics of Coal Measure strata (Fig. 4.2). The largest, some 3 ha in extent, consisted of 12 m of measures containing two seams, which were equivalent to seams whose true horizon at that point was some 45 m lower. These erratics, lying on 3 m of boulder clay and covered by a thinner skin, would appear to have been transported by glacial action on the fireclay floor of the 'lower' seam at least a kilometre and at the same time physically lifted some 45 m. Several other occurrences of rafting have been located elsewhere in the coalfield, though on a smaller scale.

The oxidation of coal can affect the volume of reserves beneath the Permo–Carboniferous unconformity. This unconformity has been exposed in several working sites and although the depth of oxidation (which averages about 15 m) may vary, there is always a sharp cut-off point between a horizon where there is a full thickness of good coal and a level perhaps only 2 m above where the coal has been almost completely weathered away with only a few traces remaining (Fig. 4.3).

In the central coalfields of England several sites have been

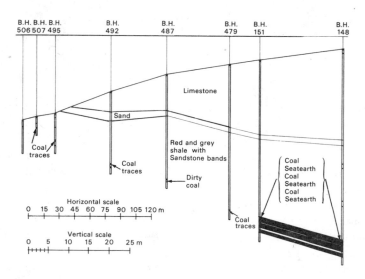

Fig. 4.3. Weathering beneath Permo-Carboniferous unconformity in coal workings, Yorkshire.

74

worked which have shown excellent examples of the effects of
tectonic disturbance of partially consolidated deposits in the form
of flow slides, local apparent 'unconformities', and discontinuities
in the succession. On some sites there was clear evidence of considerable movement of the partly consolidated deposits with steep-
sided troughs and almost horizontal floors; in others contortion
was more widespread, again over a fairly horizontal pavement. The
measures beneath the disturbance pavement were remarkably undisturbed.

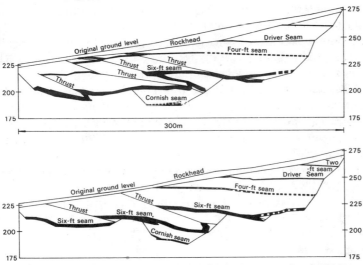

Fig. 4.4. Contortion and thrusting in Coal Measures, South Wales.

Of particular significance in South Wales are complex structures
resulting from the deformation of incompetent strata. Detailed
records show that in certain extensive parts of the coalfield differential movement of more competent strata has produced, in the
intervening beds of the productive Coal Measures, structures
revealing a full sequence from undulating beds and sharp monoclinal folds to recumbent folds and overthrust slabs, as well as
complex imbricate structures which have stacked up alternating
layers of roof, coal, and floor (Fig. 4.4). These deformations are
generally localized to zones some 50 m thick, which may transgress
the general succession and in which stresses have been accom-
modated by folding and thrusting. In many cases the behaviour of
the coal under these conditions has been such that the seams have

been deformed into fantastic shapes. In the Pembrokeshire Coalfield in the midst of areas largely barren of coal, sudden 'slatches' of thick coal are found with no visible continuity either to surface or to depth.

10. Reserves

When opencast coal production was started in Britain in 1942 it was generally believed that the total production would not exceed a few tens of millions of tons; nevertheless, by 1976 production had exceeded 310m tons.

Attention has been drawn above to the fact that the most important factor in opencast work is the ratio of overburden to coal. Considering that the figure accepted was about 5:1 in the 1940s and it now averages something in excess of 15:1, the origin of this increased tonnage will be appreciated. The relationship between tonnage and ratio is obviously not linear except in the very simplest cases, for once the ratio is increased it affects not only those seams already considered but also underlying seams which previously had been 'out of ratio'. Where then does this end? With increasingly large draglines and with improved efficiency will the ratio, and therefore the tonnage, go up and up? Remembering that opencast coal extraction is not a labour-intensive industry, will it eventually reach the stage where opencast can be a major producer of coal? These questions are unanswerable, but noting how the pendulum swings for and against the coal industry, it would be unwise to guess. Of more significance perhaps is the fact that at current ratios British opencast reserves total several hundred million tons: more than sufficient for the industry to be a potential major producer of coal for the next few decades.

References

ARMSTRONG, G., and CLARKE, A. M. (1970). Exploration and exploitation in British coalfields. *C.r. 6th Congr. int. Carb. Strat. Geol.* **2**, 417–29.

ARSCOTT, R. L., and HACKETT, P. (1969). The effect of geological features on the occurrence of gas outburst fractures in the East Midlands coalfield. *Q. J. Eng. Geol.* **2**, 89–101.

BAILEY, E. B. (1952). *Geological Survey of Great Britain.* Murby, London.

BROADHURST, F. M., SIMPSON, I. M., and WILLIAMSON, I. A. (1968). Seam splitting. *Min. Mag.* **119**, 455–63.

BROWNE, H. (1967). Coalmining: present opportunities for progress. *Trans. Instn Min. Engrs.* **126**, 825–30.

CLARKE, A. M. (1963). A contribution to the understanding of washouts, swalleys, splits and other seam variations and the amelioration of their effects on mining in South Durham. *Trans. Instn Min. Engrs.* **122**, 667–99.

—— (1966). Pay-off *v* risk assessments in planning and mining geology. *Trans. Instn Min. Engrs.* **126**, 29–42.

—— (1967). Simulation techniques in the geological assessment of engineering projects. *Proc. geol. Soc.* **1637**, 14–18.

DUFF, P. McL. D., HALLAM, A., and WALTON, E. K. (1967). *Cyclic sedimentation*. Elsevier, Amsterdam.

ELLIOTT, R. E. (1974). The mine geologist and risk reduction. *Trans. Instn Min. Engrs.* **133**, 173–83.

GOOSSENS, R. F. (1973). Coal reserves in the Selby area. *Trans. Instn Min. Engrs.* **132**, 237–44.

HOEHNE, K. (1956). Zur neubildung von quartz und koalinit in kohlen flözen. *Chemie Erde.* **18**, 235–50.

LAUGHTON, E. C. (1970). Recent application of geological methods in coal mining. *Proc. 9th Commonw. Min. metall. Congr.* **2**, 385–403.

MILLS, L. J. (1972). The utilization of resources—geology. *Trans. Instn Min. Engrs.* **131**, 235–44.

—— (1975). The Selby coalfield. *Trans. Instn Min. Engrs.* **135**, 43–51.

National Coal Board (1972). Design of mine layouts. *Working party report.*

PEARSON, G. M., and WADE, E. (1967). The physical behaviour of seat-earths. *Proc. geol. Soc.* **1637**, 24–33.

PEARSON, H. W. (1974). Coal. In *The geology and mineral resources of Yorkshire* (ed. D. H. Rayner and J. E. Hemingway), pp. 309–27. Yorkshire Geological Society, Leeds.

POOLE, E. G. (1969). The stratigraphy of the Geological Survey Apley Barn Borehole, Witney, Oxfordshire. *Bull. geol. Surv. Gt. Br.* **29**, 1–103.

RAISTRICK, A., and MARSHALL, C. E. (1939). *The nature and origin of coal and coal seams.* Edinburgh University Press.

TEICHMÜLLER, M., and R. (1968). Geological aspects of coal metamorphism. In *Coal and coal-bearing strata* (ed. D. G. Murchison and T. S. Westoll), pp. 233–67. Oliver & Boyd, Edinburgh.

TRUEMAN, A. (1954). *The coalfields of Great Britain.* Edward Arnold, London.

WEEKES, P. G. (1972). Procedure for the assessment of reserves. *Mining Department instruction.* National Coal Board.

WHITTAKER, B. N. (1974). An appraisal of strata control practice. *Trans. Instn Min. Engrs.* **134**, 9–24.

WILLIAMSON, I. A. (1967). *Coal-mining geology.* Clarendon Press, Oxford.

—— (1970). Tonsteins—their nature, origins and uses. *Min. Mag.* **122**, 119–25, 203–11.

WILSON, M. J. (1965a). The origin and geological significance of South Wales underclays. *J. sedim. Petrol.* **35**, 91–9.

—— (1965b). The underclays of the South Wales coalfield east of the Vale of Neath. *Trans. Instn Min. Engrs.* **124**, 389–403.

5. Geology in the minerals industry

1. The variety of professional demand

The geographical range of activity in the minerals industry is surpassed only by the variety of useful applications of geological knowledge. There is within geology no specialized discipline of mining geology equivalent to, say, petrology or palaeontology or structural geology; mining geology is the application of many kinds of geological knowledge and techniques to the solution of engineering problems in the extractive minerals industry. The mining geologist may find himself involved in professional tasks as different as mapping geomorphologically controlled stanniferous fossil river gravels in south-east Asia, solving complex fold patterns of layered sulphide mineralization in Pre-cambrian metamorphic rocks in southern Africa, applying sampling and statistical theory to predict the distribution and average grade of minor amounts of copper and molybdenum in volcanic 'porphyry' deposits in the U.S.A., investigating the hydrogeology of an underground mine pumping out millions of gallons of water per day in central Africa, seeking by microscopic and laboratory methods the reasons why a mineral processing plant is recovering only 75 per cent of the gold contained in an ore in Australia, or feeding the essential geological data into a team designing the optimum wall slopes for a new open-pit asbestos mine in Canada.

The ever-increasing variety of geological challenge and opportunity generated by the extractive minerals industry is characteristic of it, and the reasons for this variety may readily be appreciated if one reflects for a moment on two matters fundamental to this great primary industry. These are the wide environmental range of mineral deposits in both the geological and geographical sense, and the economic structure of the minerals industry itself.

1.1. The geological and geographical environments

A great deal of common ground exists, and is only to be expected, in the applications of geology to the field of metalliferous and industrial minerals covered in this chapter and to those covered elsewhere in this volume. One fundamental difference also is apparent, and that is in the range of geological diversity characteristic of the respective resource groups. In dealing with oil and gas, coal, cement-making materials, and sand and gravel for construction, the geological environment is broadly limited to sedimentary rocks in the upper parts of the geological column. By contrast, the wide range of useful metalliferous and industrial mineral deposits exists in all types of rocks, over the full extent of earth history, and related to every sort of structural and metamorphic condition. It is this diversity of geological environment and style which demands the application of geological knowledge of almost every kind. Even within one of the numerous specialist groups related to specific commodities the mining geologist must maintain a broad knowledge and appreciation of the controlling geochemical and geophysical environments of ore deposition. The geologist specializing in diamonds, for example, in seeking ore reserves will use his skill in geomorphology and sedimentology when delimiting the location and payability of diamondiferous marine beach gravels above and below the present sea level of south-western Africa; he will then turn to his knowledge of mantle processes, diatremes, and ultrabasic rocks in discovering and evaluating kimberlite pipes, the primary source of diamonds. The geologist specializing in nickel ores, similarly, may find himself concerned first with endogene sulphide accumulations in metamorphosed Archaean ultrabasic rocks (as in Western Australia), and then with exogene lateritic accumulations of geologically youthful age.

The variety of geological challenge generated by even one particular type of mineral deposit is strongly influenced by its geographical location. The terrain, climate, and relative isolation from major population centres all throw up engineering problems (in which applied geology can assist) against the background of general development of the country concerned. The metal mines brought into production since the early 1960s in Carboniferous rocks of the plains of central Ireland would have faced different problems had those same orebodies been developed in, say, the mountains of

Iran, not only in technical mining methods but in such matters as water supply, housing, access roads, power supply, labour, services in general, and environmental planning and control. It should also be appreciated that in a small and highly developed country such as Britain not only are specialized services obtainable relatively easily from the geological population, but also the most isolated mining geologist is never far from professional colleagues, libraries, and other contacts. At the other extreme the mining geologist who serves a remote mine in a poorly developed country not only lacks ready professional intercourse but is commonly called upon as the only available 'expert' to tackle all manner of geological problems as a geological Jack-of-all-trades.

1.2. The structure of the minerals industry

The cycle of discovery, exploration, development, exploitation, and exhaustion of an orebody was in past times usually initiated by the traditional prospector followed up by a mining company whose existence, activities, and fortunes were commonly based upon a single orebody of restricted life, and which made limited use of geology and geologists. The recent past has witnessed transition and great change in the structure of the minerals industry in response to a number of important trends in the contemporary world, and with that change there have come increasing challenges to practical geology.

(a) The modern industrialized world depends upon mineral raw materials, and the demand grows exponentially, driven faster even than world population growth by the multiplying effect of the rising overall per capita consumption of minerals which results from higher living standards (Fig. 5.1). This means that greater tonnages of ore must be mined every year by an expanding minerals industry, and an increasing rate of discovery and exploration of new deposits is required to replace those being exhausted.

(b) While reserves of high-grade and relatively accessible ores have been worked out or surpassed by demand, technological developments in bulk mining and mineral processing methods have made possible the low-cost extraction by large-scale open-pit methods of deposits of progressively lower grade. Economic considerations also have called for higher extraction rates from mines. All this has led to the development of bigger and more competent companies with engineering and financing capabilities appropriate to capital-intensive new mining ventures costing scores or hundreds of millions of pounds each.

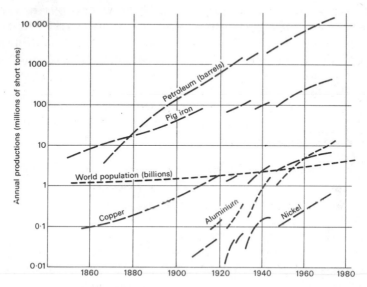

Fig. 5.1. The growth of demand for selected metals in relation to the 'population explosion'. Petroleum illustrates a major fuel, pig iron a basic industrial metal, copper a traditional base metal, and aluminium and nickel the newer metals.

(c) Technological developments in prospecting techniques have led to discoveries in terrain formerly too difficult to prospect (e.g. the discovery by geochemical methods of the important Panguna porphyry copper–gold deposit in the dense tropical forests of Bougainville). One result has been mining development in areas or countries formerly lacking a mineral industry.

(d) Environmental planning and control is receiving much-needed attention, and society is realizing that good housekeeping in the mining industry is both desirable and possible provided the cost is met in the increased prime price of mineral commodities.

(c) Because of the non-uniform distribution of most mineral resources across the face of the globe, mining and trade in minerals is not only essentially international in character but also of strategic importance politically and economically. Many governments are now taking more direct interest and control in domestic mineral exploration and mining ventures in addition to influencing developments through laws and taxes. Nationalization, or state participation with the private industry sector, is becoming increasingly common, particularly in developing countries such as Zambia and Chile, whose economies are heavily dependent upon non-renewable mineral resources. In general, a marked overlap in activity between governments and private industry is developing, as shown in Fig. 5.2.

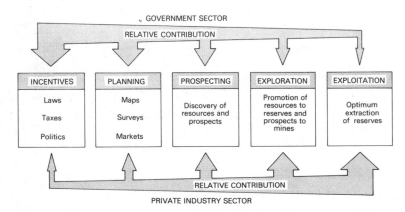

Fig. 5.2. The interplay between the state and private sectors in the minerals industry.

The contemporary structure of the minerals exploration and production industry as it affects the use of geology and employment of geologists thus incorporates the following sectors, acting either alone or in various combinations:

—mining companies ranging in size from very small operations to very big integrated corporations with international activities;

—government-controlled bodies concerned directly with mineral resources by way of one or more of the functions of planning and control, discovery and exploration, mining and production, and services (such as geological surveys, and research and productivity councils);

—specialist services and consultancy groups offering management, technical surveys, and advice to both industrial companies and government bodies.

2. Main lines of activity in mining geology

Since every individual ore deposit is both finite and non-renewable, the continuing life of the extractive minerals industry depends upon meeting demand and replacing mined-out reserves by discovery of new mineral deposits and by extending ore reserves at existing mines. The rate of exhaustion or life of a deposit in turn depends not only on the production rate of ore per annum but also in part on the working efficiency which determines what proportion of the total valuable mineral or metal *in situ* can be extracted profitably and what proportion will be left behind as uneconomic or irrecoverable. The contribution and service from geology plays its part in this industrial cycle in four main overlapping ways which are tending to divide into specialized professional fields.

At one pole is the discovery of new mineralized fields or deposits through prospecting, which makes use primarily of geological knowledge and theory in combination with search techniques such as applied geochemistry and applied geophysics. At the other pole is the application of geology to an operating mine, where the objective is to provide services which tend to optimize the overall production engineering process (and profitability) by clean mining, lowering unit production costs and raising product quality. The senior economic geologist will have experience at both poles, for it is precisely that combination of professional experience and skill in imaginative prospecting and practical daily production which is valuable in the central field of overlap, namely the detailed exploration and assessment of a mineral prospect with a view to its exploitation. The objective is to identify and quantify all the geological factors which could affect the vital forecast of profit or loss in a future possible mining operation. The fourth field covers geological services for civil engineering purposes on mines and their associated communities. Services concerned with water supplies, construction materials and sites, and many others such as stability of natural and artificial slopes, all overlap heavily with the field of modern engineering geology. In all these fields the work of the mining industry geologist is predictive, and in most cases it is quantitative and detailed as well.

3. Illustrative case-history of Kilembe Mine, Uganda

The points made above may be illustrated by reference to the case-history of a single mine.

Geology in the minerals industry

Kilembe mine is a medium-size operation remotely situated in the eastern foothills of the Ruwenzori mountains of Uganda. It has been producing approximately 2 per cent copper ore at about the million ton per annum level since production commenced in 1956. Geologists have played their part at every stage of the history of Kilembe, which has always been a 'difficult' mine.

3.1. Government contribution

The original discovery was made about 1920 by geologists and prospectors of the Uganda Geological Survey, by the traditional method of following mineralized boulders up the Namwamba River to the source outcrops. The government sector has continued to play its part at intervals ever since, in several ways. Advised by economic geologists employed by government and U.N. agencies, and by independent consultants, government development corporations have held part of the equity capital of the mining company since production began, and have participated in its fortunes as shareholders. The tax arrangements negotiated in the early 1950s provided one of the essential incentives in the decision to capitalize the mine. Various discriminatory tax regulations have followed at intervals over the years, such as export tax based on copper prices, and these have at times created disincentives to the risk financing of technical development programmes on which the ultimate life of the mine depends through exploration and definition of further ore reserves. (Governments would do well to seek professional geological advice on the effects of their proposed policies on the mining industry and individual mines.) More direct contributions to industrial geology at Kilembe have been made over the years by the Geological Survey through continuing co-operation in regional mapping, and in providing at intervals specialist laboratory services and field geophysical surveys. With the increasing and proper interest of governments in mineral resource development, there is an expanding need in the state sector for economic geologists and mineral economists trained to understand the problems of the mining industry.

3.2. The exploration phase

The Kilembe prospect was brought through the exploration stage by two companies. Tanganyika Concessions Ltd. abandoned the prospect as uneconomic about 1934 after several years of work owing to the low ruling price of copper. After the Second World War, Frobisher Limited revived and extended the work of exploration to the point of decision to capitalize a mining operation in 1952, followed by a construction phase resulting in production from 1956. Both companies conducted geological mapping, geophysical surveys, underground exploration, diamond drilling, and sampling campaigns, the Frobisher work building upon that of T.C.L. and naturally taking exploration through a much bigger programme to the stage of completion.

The decision to capitalize a new mine normally calls for risk capital in the range of several millions to several hundreds of millions of pounds and emerges from a feasibility study which projects an economic outcome on the basis of many factors. Among these are forecasts of variable product demand and price, taxes, infrastructure costs, etc., but the basic and therefore most vital factors are the unchangeable natural conditions of the orebody. Most notable are its size, shape, and internal structure, the distribution, variability, and average grade of its valuable content, its physical amenability to mining methods of varying kind and cost, and its mineralogical amenability to treatment and product quality. The definition or prediction of these vital basic factors within sufficiently narrow limits of certainty are summed up in the deceptively simple term 'ore reserves'—that portion of the mineralization which can be mined, treated, and sold at a profit.

The geological factors were difficult to establish at Kilembe because of its situation in mountainous country (difference of relief of 1000 m) covered with dense vegetation (mainly 'elephant grass' about 3 m high), and the complex Pre-Cambrian metamorphic geology (Fig. 5.3). It became apparent from a sequence of trenching, drilling, and underground workings that the surface showings of ferruginous gossans and oxidized copper minerals which were found scattered over the hills were derived from a primary mineralization of pyrite–chalcopyrite which occurred as disseminations, fracture fillings, and irregular masses in a layered metamorphic schist sequence. The ore was observed to be generally tabular and conformable with particular host rocks which were intensely folded, faulted, and sheared, and associated with masses of alaskite pegmatites. The horizontal widths of ore varied from a metre or two to over 70 m; attitudes varied from flat to vertical with dips and strikes in all directions; and there were great variations in intensity

Fig. 5.3. (a), View south-west across the Namwamba River valley 1956, showing the partly completed waste stripping from the Eastern Deposit open cut, and the newly constructed mill at elevation 4500 ft. The Bukangama ridge at nearly 7000-ft elevation forms the skyline. (b) View south-west from the Northern Deposit across the Nyalusegi Creek towards the Bukangama deposit, 1966, to the right of the view in (a). The adits can be discerned by waste rock tips. On the skyline is the hydraulic mining operation to provide sandfill for the stopes within the mountain below.

(b)

87

of mineralization and metal content. The physical conditions brought gloom to mining engineers and the financial viability was judged to be very marginal at a minimum target ore reserve of 10 million tons at the average indicated grade of about 2 per cent copper and 0.2 per cent cobalt. The greatest contribution from geology came from the detailed mapping and interpretation of surface, underground, and drill-core exposures at scales between 1/5000 and 1/250 combined with a flexible mental approach to the problems of

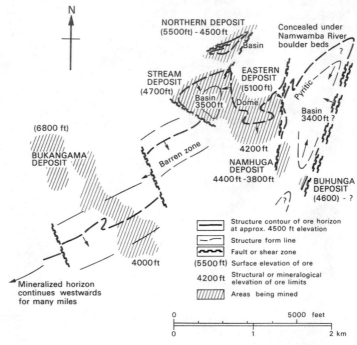

Fig. 5.4. Highly simplified structural sketch map of the Kilembe deposits.

ore genesis. In 1950 a logical stratigraphical sequence of metavolcanic and metasedimentary rocks was identified and it was recognized that these rocks had been twice metamorphosed, folded and cross-folded into complex domes and basins broken by faults and shear zones, and accompanied by metasomatic phenomena, sweat veins, and pegmatites (Fig. 5.4). The earlier view that the ore was epigenetically controlled and localized by faults, shear zones, and pegmatites was replaced by the view that the ore was syngenetically controlled and stratigraphically localized with extensive metamorphic remobilization in detail. This change of concept was immediately put to the test, with favourable results, in siting drillholes for the extension of ore, and so provided the essential confidence in the long-term possibilities of the mine in terms of tonnage. For large and rich mineral deposits the favourable combination of tonnage and grade leaves the economics of mining in no doubt, but in economically marginal deposits (Fig. 5.5) such as

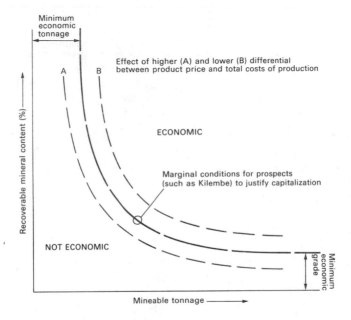

Fig. 5.5. The general relationship between minimum tonnage and grade for economic viability of mineral deposits.

Kilembe the geologist's approach to ore controls has a vital bearing on the cost and effectiveness of exploration and on the view taken of the future potential of the deposit.

Despite the professional satisfaction and practical benefits derived from a reasonably accurate discernment of the factors controlling ore deposition, the geologist is not paid during the busy and costly exploration phase to spend his time contemplating the fascination of rival theories of ore genesis. His practical work of mapping, drill logging, and sampling forms the critical foundations upon which the engineering team must build the production feasibility study. His success will be measured by his degree of integration into that team and the clarity with which he communicates his data and their significance to his non-geological colleagues. The work at Kilembe included the following projects typical of the exploration phase.

An overall geological and topographic map was built up to provide the basis of planning not only internal mine layout but the disposition of surface plant, housing, tailings dumps, etc. relative to future possible mining activity.

Ore reserve estimates were prepared and up-dated at intervals on the basis of detailed geological plans, cross-sections, longitudinal projections, and structure-contour maps which recorded the factual results of underground and surface geological mapping, core logging, and sampling, plus interpretation of ore outlines between exposures. The detailed three-dimensional picture of the orebody and its environmental rocks was displayed also on models. The approach at Kilembe was to test all suspected areas of ore with relatively widely spaced drill-holes to provide an overall picture of ore in the 'possible' category. Two accessible areas were also explored in great detail by underground drives, crosscuts, and raises, plus underground diamond drilling, to provide the assurance of detail, continuity and mineability appropriate to the 'probable' and 'proved' categories of ore reserve. It was a geological responsibility to ensure that these areas of detailed work were in fact representative.

An ore reserve estimate generally consists of a statement of tonnage and valuable content correct within certain limits. The tonnage estimate depends heavily on the geologist's detailed interpretation of ore limits to define a volume of rock which can be transformed to tonnage via bulk density. Density will vary locally in most mineral deposits with grade of ore, and requires careful treatment. At Kilembe density variations are associated with increase of sulphide content, but not directly with copper content since concentrations of pyrite or pyrrhotite might be barren of copper though containing up to 2 per cent cobalt in solid solution. The grade aspect is more difficult, and rests upon the results of a sampling campaign to reveal the distribution and variability within the orebody of the valuable minerals. From the sampling or assay data the average grade of suitable blocks or subdivisions of the orebody can be estimated by various weighting procedures or statistical methods. Such procedures are amenable to computer methods, but the overriding factor of importance is the recognition of variability in terms of geology.

The sampling campaign at Kilembe was critical because the decision to place these marginal ores in production was heavily dependent on the accurate establishment of the average ore grade. The task was accomplished very successfully, as proved by mining results after 1956, by using several methods to check one another, all under strict geological control. Systematic samples were obtained from split diamond drill cores, from underground exploratory workings by channel sampling and chip sampling ore *in situ*, and by grab sampling of broken ore from the same workings. A pilot plant was then used to crush, grind, and treat the ore from a number of crosscuts excavated along the line of earlier horizontal drill-holes, so providing a bulk sample reference value against which the several sampling methods could be directly compared for bias and mean deviation. Bulk sampling through a pilot plant is commonly used to test the reliability of sampling by drill-holes, for instance in low-grade 'porphyry' deposits, where a systematic bias error could have very serious consequences. In some cases of erratic mineralization such as 'spotty' gold deposits or pegmatites, bulk sampling is the only feasible way of obtaining samples large enough to be meaningful. There is an obvious difficulty in deciding on the optimum number and spacing of drill-holes or crosscuts for sampling an orebody, and on their pattern and direction. Statistical methods are rapidly being developed which help to replace the empirical approach to the difficult problems of sampling orebodies. The reduction and preparation of samples for assay are yet another source of error and possible bias. At Kilembe this work also was under charge of the geologists rather than the assay office, for it is the geologist who in estimating the ore reserves must appreciate the overall degree of error compounded from the errors inherent in each step in the process from the taking of an individual rock sample to the estimated average grade of a complete orebody.

Mineralogical characteristics were established for Kilembe ore through careful megascopic mapping and laboratory examinations, and zonal patterns were looked for over the known extent of the orebody. It was quickly established that the primary sulphide minerals were essentially chalcopyrite as the only copper mineral, with two varieties of pyrite and pyrrhotite characterized by up to 2 per cent cobalt in solid solution, and a cobalt nickel sulphide, 'linnaeite'. The microscopic intergrowths and grain-size range indicated that flotation processes would work well on relatively coarsely ground ore. The mineralogical conclusions were borne out by flotation tests in the pilot plant and copper and cobalt concentrates were produced at very acceptable efficiencies and costs. At the same time areas of oxidized ores were defined, and also an important zone of supergene enriched chalcocitic copper ore was located. The mineralogical characteristics and tonnages of these ore types were shown to warrant a separate mineral concentrating plant of planned limited life. Mineralogical studies are a most valuable geological service during the proving stage of a mine and in the design of the mineral extraction plant. At Gortdrum mine in Ireland, for instance, the first electron-probe study of the ore revealed the presence of economically important amounts of mercury, previously unsuspected from sampling and assaying procedures, in the form of mercurial tennantite.

The design of mining methods involved geologists in detailed studies with mining engineers to evaluate all the factors which affected the choice of mining methods and so related to mining costs estimated for the economic feasibility study. Such factors include the detailed shape of the orebody from place to place, the manner in which mineralization either ends suddenly or fades gradually at contacts and so affects estimates of wallrock dilution and stope limits, the joint patterns and physical condition of the ore and wallrocks, which affect stoping methods and costs, and other related parameters, all derived from the detailed geological mapping of underground workings and logging of drill cores.

3.3. The construction period

After a feasibility study based on detailed exploration work has concluded that a mining operation will be worth while, a construction period, normally of one to three years duration, is required before production can commence. This is a financially critical period, because capital is spent at a high rate before cash from product sale flows back into the enterprise. Delays in the scheduled start-up date, especially in times of high interest rates and rising costs, can be extremely expensive. Although the period is clearly the responsibility of the construction engineer, the geologist still has important services to offer in preparatory mining works and in civil engineering works.

Kilembe was typical of mining developments in remote areas of low population, and the present-day status of the Kilembe–Kasese township complex as the third largest in Uganda is due to the primary and secondary employment opportunities at and near the mine and the 200-mile westward extension of the railway from Kampala for which the mine provided the final economic incentive. A modern mining industrial complex now exists, together with townships, roads, and power and water supplies, in the Namwamba Valley where elephant and buffalo were formerly more numerous than people.

In a highly developed country it is difficult to appreciate that the first step at places like Kilembe is to create a topographic map, usually with the assistance of air photo cover and photogrammetry. The geological map which follows, normally at a scale near 1/10 000, forms the basis of planning. Obviously no important installations should be constructed over any areas likely to be undermined, and the geologist's forecast of long-range ore possibilities is important but not always easy. There seems to be a natural tendency to build as close as possible to the mineralized centre of interest, with the best sites on the hanging wall of the orebody! At Butte, Montana, part of the town has been abandoned to make way for open-pit mining of material of too low a grade to have been thought of years ago as future ore. At Thetford Mines, Quebec, extensive adjustment of railway routing and township facilities was necessary in the early 1950s when it became apparent that the full extent of the asbestos orebodies had not originally been realized. The important and early realization at Kilembe that the primary control of ore distribution is stratigraphical, and a general appreciation of the structural geology, helped in planning wisely in a mountain area where space was at a premium for the siting of housing and community activities, offices, workshops, the mill, and tailings disposal sites. Nevertheless when it was discovered after some years that the mineralized MK horizon, displaced some four miles from the main mine area, ran past the club house and down number one fairway of the golf course, the geological members

93

were relieved to find that the mineralization was in that locality very much subeconomic!

Despite the vital planning aspect linked to geological forecasts of where mining may be done in future years, the operational geologist is during the construction phase more occupied in the practical day-to-day demands on his skill in engineering geology. The work done at Kilembe is illustrative of the flexibility demanded of the mining geologist unless specialists are brought in.

Supplies of sand, clay, and rock aggregate had to be located to meet construction demands in quantity and quality in an area where the most sophisticated existing structures were of mud-brick and thatch. A brick- and tile-making plant was established on the site of a good clay deposit discovered by prospecting the Kasese flats below the mountain front between areas of mountain stream debouchment. Suitable sand deposits were similarly selected, after prospecting and testing for quality and quantity relative to cost of transport to the point of use. It might be thought that rock aggregate is no problem at a developing mine, but this is not always the case because of the specifications for aggregate for structural concrete. At Kilembe the waste rock from underground exploration was suitable for bulk road fill and many other purposes, but the admixture of sulphide-bearing rocks, inter alia, made it unsuitable for structural concrete. Deposits of rounded river boulders were heterogeneous and difficult to crush, and in the end a special quarry site was selected in homogeneous gneiss of suitable physical characteristics.

The site for the mill was selected with care on the lower hill slopes above the valley floor at a site suitably close to and below the main tramming adit level from the mine. The site area was traversed by one of the many shear zones at Kilembe and also by a wide dolerite dyke, younger than any tectonic events except the latest stage of faulting, but rotted near the surface by tropical weathering. A typical geotechnical site investigation was conducted here in conjunction with the civil engineering team to achieve the optimum design layout, foundation conditions, and stability for the heavy crushing, grinding, and flotation plant. Landslips are common on the youthful mountain slopes at Kilembe, which are characterized by tropical weathering and seasonally heavy rainfall. Unstable conditions are watched for, and care is taken with drainage on important sites.

Another example of geological mapping control for civil works at Kilembe concerned the power supply. Hydro-electric power was to be generated in the near-by but larger Mobuku River valley by conducting some 200 cusecs of river water to a penstock via an open flume supported on trestles. The flume line traversed along the contours some miles of hill slopes, all of which was carefully mapped to establish ground conditions, unstable areas, etc. for engineering design control and to assist in such decisions as whether to tunnel through or traverse round sharp spurs.

The source of water supplies for domestic and industrial use was a relatively simple matter at Kilembe because of the adequate supplies in mountain streams at elevations suitable for gravity supply. The main domestic supply was selected from a drainage basin shown by geological mapping as unlikely to be polluted by any mining activity or future dense population growth. In many new mine projects in arid areas by contrast the problems of water supply may be critical. The mining geologist finds in this field one of the many challenges to his versatility in the practical use of geological knowledge, and to his common sense in knowing if, and when, to recommend bringing in a specialist hydrogeologist.

Pollution and environmental quality control are also matters in which geologists become involved in the construction stage, with a wide variety of problems. At Kilembe, trace-element surveys were conducted in the early 1950s to establish background levels of copper entering the shallow Lake George, which supported a local fishing industry. A sulphide-roasting plant (for cobalt production) was also projected at Kasese, the railhead, and the height of the smoke stack was important to avoid future contamination of projected cotton-growing land by sulphurous smoke. This was determined by a meteorological study in which geologists (for lack of more suitable local scientists!) participated. The whole subject of environmental quality control has in recent years been brought more sharply into focus, and at all new mines geologists are certain to be involved, particularly in the pre-production stage when natural background geochemical levels of heavy metals should be established in waters and soils.

The second aspect of geological work during the construction period concerns pre-production mine development, involving mining geology in its own specialized field. During this period there must be accomplished all the mine preparatory work for actual production, such as waste stripping for an open pit, or shaft sinking and underground development to prepare blocks of ore for stoping in readiness for the commissioning of the ore-treatment plant. A good recent account of one aspect of this kind of work in Britain concerns the geological contribution to the detailed engineering planning of the shafts at Boulby for deep underground mining of Permian potash salts (Woods 1973). That account by the resident geologist illustrates well the quantitative nature of mining geology and the close teamwork needed with mining engineers. It also shows in detail the co-ordination of geology, geophysics, drilling technology, and testing procedures for the 3947-ft deep borehole which provided the pilot data vital to the major engineering task of deep shaft sinking in difficult ground conditions. This planning and control aspect is important at all mining operations in the interests of efficiency and profitability. At Kilembe, as in other underground mining operations, the main haulageways, ore transfer passes, and other essential long-term works all had to be carefully positioned in relation to the geometry of the orebody; they had furthermore to be driven in the best available ground to avoid premature failure or expensive artificial support. All this depended on detailed geological mapping and forward projection. At Kilembe two small open pits were used to exploit parts of the orebodies where they were thickened close to the surface by folding and faulting. Geological prediction of the ore outlines was required in order to design the pits, and during the construction period the barren overburden was stripped by contractors. The contract price struck rested mainly on two factors in which geology played a large part. First, the volume and disposition of barren rock to be removed under contract depended entirely on the geological prediction in detail of the shape and position of the irregular upper ore limits relative to the topography. Second, the cost per unit volume depended on whether the material could be removed by scraping, ripping, or drilling and blasting. The geologist's prediction of the disposition of various rock-types and their physical condition, based mainly on interpretation of drill cores, was again a prime factor. The practical usefulness of geology to the waste-stripping contractor depends, as it does to all non-geologists, on effective communication of geological opinion. The contractor may be excused his disappointment if he learns only that the rock to be removed is '... a coarse-grained metasediment showing evidence of amphibolite-grade metamorphism with later retrograde metamorphism, correlatable with two phases of folding with axes crossing at high angles'. His attention is more likely to be captured by '... hard, gritty, and abrasive rock, but so heavily fractured that ripping seems feasible ...'. The angle of slope of the pit walls is also of concern. Too flat a slope by even a little means money wasted, and too steep a slope means possible instability and premature failure. The optimum angle will obviously vary with the geological conditions, and the earlier these can be defined the better. It is normally either difficult or costly, or both, to start again to cut back the lip of a pit which has already proceeded to some depth. At Kilembe a considerable saving was effected in the Eastern Deposit open cut by early recognition that the north wall was formed by steeply dipping ore underlain by a massive alaskite pegmatite which should (and did) stand at a high angle. With the increasing use of open-pit mining, rock-slope engineering is rapidly developing refinements, and in this the geologist plays a very important part, employing special mapping techniques to emphasize the critical features such as discontinuities at various scales, and the groundwater conditions.

3.4. The producing mine

The producing mine is the core of the minerals industry, and the geologist plays his part in the dual functions of exploration and conservation. Exploration (as distinct from prospecting) is the work of defining the ultimate worth and limits of an ore deposit through investigation of any direct or indirect indication of mineralization. Dictionary definitions of 'conserve' include the appropriate phrases 'to keep from waste or loss, to manage wisely'. The overall life and profitability of a mining operation rests first upon discovering and defining all the ore available, and secondly upon wasting as little as possible of it through the inevitable losses involved in the processes of mining and treatment. The first is clearly a geological responsibility; the second an engineering responsibility in which geological service plays its part.

The maintenance of ore reserves. Miners do not enjoy the privilege of designing a finished article, since the full extent and shape of the economic mineralization cannot be known in detail until it has been mined out—and perhaps not even then. When a new mine is brought into production, only enough ore will have been proved and indicated to provide a reasonable assurance to the owners that their risk capital will be returned with interest over the life of those ore reserves. The greater part of the exploration of the orebody in detail is a process continuing over the life of the mine, so that by the advance of mine exploratory drilling and development openings each year enough ore is successively promoted through the classes of 'geologically inferred' to 'possible', 'probable' and 'proved for mining', to balance the amount removed by mining. At Kilembe, for instance, expenditure on the exploration phase was stopped when the proved and probable ore reserve amounted to the minimum economic requirement of approximately 10 million tons at an estimated average grade of about 2 per cent copper plus about 0·2 per cent cobalt. Mining since 1956 has been at rates between 0·8 million and 1·2 million tons per annum—totalling some 18 million tons—but the current proved and probable ore reserve is still near the same general level. Evolution of this sort is normal in mines, but it must not be supposed for that reason that maintenance of ore reserves is an easy task. Especially in orebodies of complex shape and variable quality, the geologist must seek out the factors controlling ore concentrations, he must map, predict, test, record, and report his findings. His objective is to provide the data for management decisions to extend the life of the mine at the optimum scale and sequence of operations. The mine geologist, in common with many other industrial geologists, enjoys the rare satisfaction of knowing that his theories and quantitative predictions are constantly being put to the practical test of the exploration drill and the miner's excavation.

Geological services to operational efficiency. Mining operations are complex and cannot be 100 per cent efficient. The important effect of any reduction in losses, or improvements in unit costs and efficiencies, is particularly well recognized in the engineering control of those operations which are of only marginal profitability. Geologists form a part of the engineering team, and the value of geological services to the departments of planning, mining, mineral processing, and civil engineering depends essentially on their predictive and quantitative nature.

Services to the engineering department. Work of the sort done during the construction period continues throughout the life of a mine, and skills in engineering geology and geotechnical mapping may be called upon in many ways. Prediction of geological hazards is a common call on services. Examples range from road routes at Kilembe which might traverse unstable shear zones or landslide areas on steep mountainsides, to the difficult prediction on certain western Witwatersrand gold mines of the sites of near-surface natural caverns in dewatered Transvaal Dolomite which may collapse with disastrous results for built-up areas. One of the long-established concerns of the mining geologist has been 'ground control', and there is a growing demand for geological skills in this field shared by mining engineers, civil engineers, and geologists concerned with rock mechanics, ground stability and behaviour in underground mines, subsidence due to undermining, and the design of open-pit slopes.

Services to mineral processing plants. Throughout the life of a mine improvements in efficiency and mineral recovery may be brought about not only by technology but by awareness of changing geological conditions. At Kilembe the concentrator for oxidized ore was discontinued when it was geologically certain that no economic tonnage of such ore remained, and geological attention was turned to bacterially assisted leaching of remnant ore *in situ* in worked-out portions of the mine. By mapping mineralogical and geochemical trends within the orebody, geologists made the concentrator and smelter aware of expected long- and short-term variations in copper–cobalt ratios and in the proportions of pyrite, pyrrhotite, linnaeite, and chalcopyrite which affected the optimum practice in differential flotation. At copper mines with more variable mineralogy than Kilembe, such as Roan Antelope in Zambia, knowledge of the now well-known lateral mineralogical zonal pattern was important. At times when mining was limited to bornite and chalcocite rich ores, iron-rich sulphide concentrate of pyrite or chalcopyrite had to be brought in to balance the smelter feed. This could easily be planned in advance when it was known that the mineralogical basis of, say, a 3 per cent copper ore was chalcocite rather than chalcopyrite. Similar benefits in advance planning and control are derived from geochemical knowledge of the distribution trends throughout the orebody of impurities or by-product elements.

From time to time at Kilembe, transmitted and reflected light microscopic examination of mill feed, tails, or middlings was

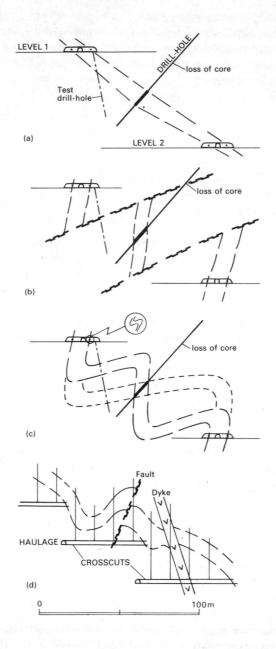

helpful, and microscopic examination was also made of smelter slag as part of a study of the feasibility of retreatment. Some large mining companies maintain research laboratories for solving problems and seeking improved methods of ore treatment, and the mineralogist plays an important part.

Services to the mining department. The day-to-day work of the geologist is thoroughly integrated with that of the mining engineer in the efficient production of ore, in the following main ways.

Outlining the boundaries of ore to be mined is essentially a predictive process of interpolation between exposures provided normally by drilling, and to a degree of certainty appropriate to the mining method being used, but always in the most quantitative way possible, taking into consideration all the geological factors known to typify the orebody. The importance of correct geological interpretation is illustrated in the simplified diagrams in Fig. 5.6, where straight-line interpolation is most misleading. The outline of an orebody is seldom a simple matter, for it depends not only on structure but on the cut-off grade of mineralization between (profitable) ore and (non-profitable) waste, with demarcation of marginal material. The correct prediction of ore outlines, including faults, folds, or other irregularities is vitally important to the mining engineer in designing his ore-extraction system and its costs, and mining geologists must not only be expert at such geological techniques as structure contouring but they must also be thoroughly

---

Fig. 5.6. Simplified vertical transverse sections.

(a), (b), and (c) show different interpretations of the ore structure between two underground levels and a drill-hole. (a) simply joins assay boundaries by straight lines; (b) takes into account the measured dips in the crosscuts, the intersection angle in the drill core, and interprets loss of core as evidence of faulting; (c) considers in addition the observed minor structures, with two fold solutions. These possibilities could be tested by a drill-hole as shown from the upper level. The correct solution affects both the estimated ore reserve between levels and the layout of mine development openings to extract the ore.

(d) Primary footwall development at 80-ft level intervals and vertical diamond drill-holes at 40-ft intervals to delineate ore ahead of cut-and-fill stoping at Kilembe. In practice the cross-sections (every 80 ft along strike) would show much more geological detail. Crosscuts do not traverse the orebody to avoid creating bad ground conditions. (After Bird 1968).

acquainted with the techniques and economics of the appropriate mining methods.

Grade control or quality control of production has the objective of meeting the target production figures called by management (through a planning group in which the geologist would be included) for set periods such as a day, month, or year. The target at Kilembe for instance might be 90 000 tons of ore at average grade 1·80 per cent copper for a month, compounded from a dozen or more widely separated stoping areas of varying metal grade and productive capacity. The correct blend is important to meet the logically planned and balanced depletion of ore reserves from different parts of the mines and to send to the mineral processing plant a flow of ore which fluctuates as little as possible from day to day, since its operating efficiency depends on a uniform quality of feed. On some mines the product must also meet specifications subject to bonus or penalty on sale. In this field of grade control geology and the geologist have become increasingly useful. At Kilembe each sectional underground geologist maintains detailed mapping in all his stopes, working closely with the production engineers to ensure that ore is broken to the correct walls, that changes in predicted ore outlines are identified, and that payable material is not transported to the waste dump nor unpayable material to the mill. The work demands knowledge of the methods and theory of sampling ore *in situ* and in broken form, and of the appropriate statistical treatment of the assay results. Production grade control in open pits is a similar exercise, where several shovels may be loading ore of varying grade from a number of sites on several benches, and output from each working place must be controlled to achieve the desired blend. At the Cassiar open pit asbestos mine for instance, the geologist predicts the quantity and quality of chrysotile fibre in zones on a grade control map by logging exploratory drill cores and cuttings from production blast holes. There is thus available a prediction of the grade of material expected from each blast on each pit bench and the progress of the pit can be planned to achieve production grade targets from day to day and also in the longer term. Fig. 5.7 shows the result of establishing grade control practices at an iron ore mine where product quality was improved and made less erratic.

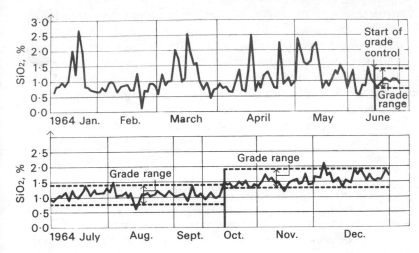

Fig. 5.7. Silica content in iron ore shipments from Lamco mine in 1964. The introduction of geological grade control resulted in closer tolerances in ore production and consequently in the constancy of the shipped product to buyers' specification. (From Gauvin 1966.)

Ground control at Kilembe has been concerned mainly with the orebody and its immediate wallrocks and their discontinuities, such as weak shear zones, gouge-faults, and fracture patterns and intensities. This is because of the underground stoping methods used in the orebody, which is broadly tabular but very variable in attitude and thickness mainly due to double folding and the presence of alaskite pegmatite masses. Experience has led to the general adoption of cut-and-fill stoping, and geological prediction of the physical state of the hanging-wall is very helpful, especially in flat-dipping zones where work must be conducted under large expanses of roof. From the early exploration days, drill log forms at Kilembe have included a special column for ground conditions in company with others for lithology, structure, and mineralization. In all mines, Kilembe included, the accurate mapping and projection of bad ground conditions such as fault and shear zones, decomposed areas, or weak rock horizons, is important to the planning of underground development work. It is especially important that major works such as shafts, underground crusher stations, and ore passes should be sited not only in positions favourable to the geometry of the orebody, but in the best available ground. The involvement of geology in engineering in this respect is obvious. Subsidence through undermining, and open-pit slope engineering have already been mentioned, and on many mines systematic studies in rock mechanics include both geologists and engineers on the team. Such studies apply, for instance, to mines employing caving methods of mining, where controlled failure of the orebody and hangingwall rocks is required. At one asbestos mine in serpentinite for instance, improved mining methods have resulted from an understanding of the detailed geology together with knowledge of a strong regional residual stress in a near-horizontal plane.

Water problems constitute one of the main hazards of underground mining, and hydrogeological studies have a prominent place in the activities of geology departments on some mines. Water under high pressure or in large volumes has caused spectacular disasters in mining, and is always expensive to control. Kilembe has confronted no really serious water problems; the only aquifers in the metamorphic rocks are shear or fracture zones of limited capacity. Nevertheless geologists have had to define through advance drilling the position and pressure of the water-bearing zones ahead of important levels and shafts into virgin areas, and give warning of probable requirements for cementation cover. The experience at Bancroft copper mine on the Zambian copperbelt is quite different, where some 70 million gallons of water per day are pumped to the surface: about 60 tons of water for each ton of ore hoisted. The early development of the mine was seriously hampered by excessive quantities of underground water, and Bancroft's large neighbour Nchanga had in fact been shut down for a time during its early development by an uncontrollable inrush of water into the main incline shaft. The major engineering problems of water control on these famous strata-bound copper mines were satisfactorily solved only when the hydrogeology of the folded sedimentary sequence was worked out on the local and regional scales. That work defined the complex system of stratigraphically controlled aquifers and aquicludes and has been described (for Bancroft mine) by Whyte and Lyall (1969). In most mining operations the need does not arise for such dramatic water-control measures, but nearly all demand an awareness by the geologist of the power and potential danger of water. Knowledge of the importance of even relatively small quantities of water under pressure in relation to tip stability, landslides, and shear zones underground and in open-pit walls, for instance, constitutes an essential part of the mining geologist's skills.

Forward planning and exploration. The geologist is particularly well placed to participate in both the short- and long-term planning which is at the core of systematic mining. He is privileged to possess the most intimate understanding of the habits and vagaries of the orebody in detail, and a superior appreciation of its possibilities of extension on the larger scale. In addition to his geological knowledge, however, his effectiveness in the planning team rests on his understanding of mining methods, ore treatment, and the cost structure of these activities.

TABLE 5.1

Simplified long-term production planning schedule to maintain
1×10^6 *tons production annually*

Area	Ore reserves remaining (tons $\times 10^6$)	Year	1	2	3	To be mined (tons $\times 10^6$) 4	5	6	7	8
1 (old)	0·7		0·4	0·2	0·1	—	—	—	—	—
2	3·3		0·4	0·5	0·6	0·6	0·6	0·4	0·2	—
3 (new)	1·7		0·2	0·3	0·3	0·3	0·3	0·2	0·1	—
4	?		—	—	—	0·1	0·1	0·3	0·4	0·4
5 (future)	?		—	—	—	—	—	0·1	0·3	0·4
6	?		—	—	—	—	—	—	—	0·2

At Kilembe a long-term ore-production schedule of the form illustrated by Table 5.1 based on ore reserve estimates set the time-scale for the many activities preparatory to ore production from a new area. As one part of the mine was progressively exhausted, so another had to be explored, ore reserves established, mine development openings put in the correct place, and all made ready for production stoping. The first producing areas at Kilembe were close to the mill and above its elevation of 4500 ft. As the productive capacity declined from the Eastern Deposit open-cut and then from the Northern Deposit, more ore had to be sought at greater depth and further afield. This resulted in a lower-level adit at 4300 ft elevation and an internal vertical shaft on the Eastern Deposit, and the opening up of the Bukangama deposit by several new adits and a 3000-m extension of the main 4500 level of the mine. Such new capital works take time (years) and money, and the geological exploration which precedes them similarly needs time and must be initiated far enough in advance of production requirements, even if a successful outcome is assumed to be likely. Two long hanging-wall crosscuts at Kilembe took over two years to complete before providing the stations for downward diamond drilling for which they were designed. Many more months of drilling followed before it could be proved that down-dip extensions in that portion of the mine were sub-economic; forward planning had to be adjusted and geologists had to revise their views on factors controlling the distribution of ore beyond the known limits. The development of Bukangama was similarly followed in due course by new extensions southwards from the Eastern Deposit after diamond drilling to follow up geological ideas based on underground mapping and surface geological, geophysical, and geochemical surveys. Each new area brings its own new challenges to the geologist; his task does not end with the discovery of new ore and its exploration and definition in terms of proved and probable ore. At Bukangama, for instance, cut-and-fill stoping was indicated, a relatively expensive mining method. It was the practice to utilize deslimed mill flotation tailings for back-filling stopes, but the Bukangama stopes would be centred 3 km from the mill and an expensive 500 m higher. A geological contribution to lower costs came from the idea that the UK4 hangingwall metaquartzites might be sufficiently weathered on part of Bukangama ridge to be useful as sand fill. This proved to be the case and by diverting water from the near-by Dungalea River a hydraulic mining operation provided relatively cheap and effective sand fill which was gravity-fed to the new stoping area.

On many mines the staff geologists become involved in prospecting and exploration outside the mine property limits, so that the scale and life of company operations can be planned in the long term. At Kilembe, with little geological hope for really large extensions of tonnage or richer ore at the mine itself, the search for additional ore deposits moved into the surrounding country even before production began in 1956, and has continued intermittently ever since. Bearing in mind that a new discovery must be considerably richer or bigger to yield the same profit as a lesser orebody close to the existing plant and other facilities, the search for outside ore near Kilembe has resulted in the view that no area offers better prospects than the close-in extensions of the mine itself. Geological opinion such as this clearly influences future company planning and priorities. As in many companies, there has been a continual interchange between the geologists concerned with the mine and those concerned with outside exploration, with mutual benefit to both the company and to the individual geologist through variety of experience and professional challenge. The exploration of the southern Ruwenzori mountains by means of air photo and ground geological mapping co-ordinated with geochemical prospecting methods, and followed up by geophysical survey and drilling in selected areas, was technically most successful. A larger-scale picture was obtained to complement the detailed mine geology, and confirmed the primary stratigraphical control of copper–cobalt mineralization over more than 40 km of strike, together with important structural data. The feedback through working at different scales on the same general problem is always useful in formulating the geological theories which guide the exploration for ore, and rewarding to the individual geologist.

Kilembe is no exception to the general rule that most mine operating companies are associates of larger company organizations which offer progression into a wider and more general field of mineral prospecting. That field is the subject of another chapter in this volume. The point may however be made again that many companies prefer their senior geologists to have experience in both general prospecting and operational mine geology in order to achieve the balance appropriate to a practical economic geologist of senior status.

4. Education and opportunity

Enough ground has been covered by way of example, however briefly, to indicate that applied geology has earned an essential place in the structure of the minerals industry. Its practitioners may be found in government service, in consulting and contracting, and in mining companies large and small the world over. Some have risen to prominence, such as company presidents, for the opportunities exist to progress within the field of management and administration as well as in several technical areas. It will also be apparent that professional competence is the product of experience and specialized industrial training built upon the basis of academic discipline and qualifications. The interrelationships between first-degree courses, specialized post-graduate courses, and industrial experience as the appropriate channels of learning, may readily be appreciated by reflecting upon the special knowledge and skills required to perform some of the varied functions of geologists within the minerals industry.

Much of the mining geologist's work, for instance, centres upon the detailed definition, sampling, and evaluation of mineralization, which calls for a combination of science, engineering, and technology within a framework of cost-consciousness. His chief tool in exploration and sampling is the drill, particularly the diamond core drill, and his knowledge of drill technology is essential to his use of drill-holes. One aspect of this is surveying the subsurface deflection commonly suffered by drill-holes so that geological information can be plotted without distortion at the correct spatial co-ordinates. Another aspect concerns the use of down-hole instrumental logging, which has not been as well developed as in the petroleum (and coal) industries because the holes are too slim for most existing instruments. Many opportunities exist for development in this special field. Existing uses have included magnetometry for interpretation of drill results in complex bodies of iron ore, and *mise-à-la-masse* surveys in sulphide orebodies, which yield geophysical data valuable for interpretation between drill-holes. Others yield data within drill-holes, such as radiometric surveys, and analysis of certain base metals such as tin and copper by portable isotope fluorescence instruments. Knowledge of sampling is essential, for systematic mining depends on accurate forecasting of the recoverable mineral values at working faces. The theory and practice of sampling mineral deposits involves the statistical analysis of data, and a field of geostatistics has developed which relates to the special statistical problems and procedures pertinent to mineral deposits. In this field also there is wide overlap between the mining geologist and the mining engineer. Much ore sampling is more highly specialized than the simplest possible case of a chemical analysis for say gold, or copper. Drill cores in kaolin, for example, may be tested for up to twenty relevant properties. Again, in stockwork-type chrysotile asbestos deposits the visual estimation of fibre value per ton of rock involves not only the overall percentage of fibre present but also its classification into lengths and qualities related to the several commercial products produced by the plant (Fig. 5.8). In that case personal bias is an obviously important factor, but knowledge of the general problems of bias error in the taking,

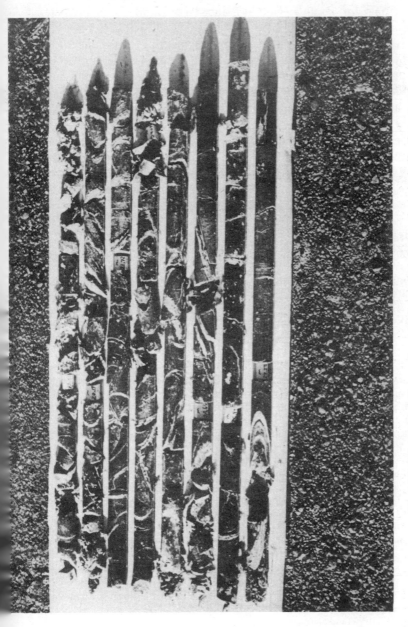

Fig. 5.8. Diamond drill core from a chrysotile asbestos deposit, indicating the difficulty of estimating the percentage and quality of fibre present by visual sampling.

preparation, and analysis (or testing) of samples is one of the essentials for mining geologists faced constantly with predictive valuation problems.

Relatively few universities offer first degrees where the special subjects appropriate to mining geology, or even an adequate lecture course in mineral deposits, are built upon a firm basis of general geology. For this reason, many graduates in geology seek research or a postgraduate course of specialized instruction either before they enter the minerals industry, or after they have gained a few years' experience. It is not the purpose of this chapter to discuss matters of education and research. Nevertheless it is appropriate to comment in conclusion that for the mining industry geologist his most pertinent and valuable educational attribute is his versatility and ability in his own special subject—geology.

References

GAUVIN, J. (1966). Open pit mining at Lamco, Republic of Liberia. *Trans. Inst. Mining & Met.* **75**, A5–A10.

WHYTE, W. J., and LYALL, R. A. (1969). Control of groundwater at Bancroft Mines Ltd., Zambia. *Proc. 9th Commonw. Min. Metall. Congr.* **2**, 173–204.

WOODS, J. E. (1973). Potash exploration in Yorkshire: Boulby mine pilot borehole. *Trans. Inst. Mining & Met.* **82**, B99–B105.

References to Kilembe

BARNES, J. W., BARBOUR, E. A., and SMIT, J. S. (1962). Kilembe copper mine Uganda. In *Stratiform copper deposits in Africa*, 1st part, (ed. J. Lombard and P. Nicolini) pp. 185–95. Association of African Geological Surveys.

BIRD, H. H. (1968). Falconbridge's copper operations in Uganda. *Can. Min. Metall. Bull.* **61**, 1075–82.

——, and KOTTLER, N. (1969). Cut and fill mining at Kilembe Mines Ltd., Uganda. *Proc. 9th. Commonw. Min. Metall. Congr.*, **1**, 171–91.

DAVIS, G. R. (1962). Results of comparative sampling methods at Kilembe Mine, Uganda, during the exploration phase. *Trans. Inst. Mining & Met.* **72**, 145–64.

—— (1963). Notes on the tectonics of Kilembe, Uganda. In *Stratiform copper deposits in Africa*, 2nd part, (ed. J. Lombard and P. Nicolini), pp. 214–15. Association of African Geological Surveys.

—— (1969). Aspects of the metamorphosed sulphide ores at Kilembe, Uganda. In *Sedimentary ores: ancient and modern* (*revised*). *Proc. 15th Inter-Univ. geol. Congress 1967* (Department of Geology, University of Leicester).

6. Mineral exploration

1. Introduction

Mineral exploration is an activity considered by many to be more of an art than a science. In fact, the work involves a knowledge of several branches of science and the application of that knowledge to the basic discipline of geology, allied with an ability to acquire, synthesize, and use information, sometimes under conditions of considerable physical discomfort. In this chapter I shall outline the thought, planning, and knowledge that go into establishing and carrying out a base metal exploration project and shall try to give an idea of the kind of life that an exploration geologist leads in the field.

In trying to exemplify a typical exploration programme, it is inevitably necessary to present an oversimplified case. No project runs a course as straight as that outlined here. One important factor is that much of the direction of a programme is related to the experience of the man handling the work and would be based on his intuition as well as his cold interpretation of facts.

A paradox for the exploration geologist is that the evolution of knowledge and technology have made his work both easier and more complex. Theories concerning the genesis of rocks and the formation of ore-bodies are continually changing or being modified with the evolution of the geological science. In addition, the increasing pace of technology in recent years has allowed scientists to develop new and far more precise analytical methods than were available previously. Advances of this type enable the geologist to formulate more accurate hypotheses about the actual mechanisms controlling the various geological processes than were previously possible by less sophisticated laboratory techniques, by observation of hand-specimens, and by field mapping. In the same way the rapid growth of electronics, computers, and other technologies has led to the development of numerous field prospecting instruments in the past ten to fifteen years.

All these factors have helped the exploration geologist in his role but have added to its complexity. His role has undergone a radical change since the time when mines were found purely by surface prospecting in areas where exposures were good and many essentially fortuitous discoveries were made.

It is becoming less and less likely that any really significant discoveries will be made solely from surface prospecting in areas which have already been explored. The exploration geologist is therefore driven to extend his search to ground which has not previously been covered by mineral explorers', or to ground where earlier searches were specific to minerals other than those for which he is searching; or to attempt to detect bodies of mineralization which lie at greater depth beneath the earth's crust than can be detected by surface observations. In addition, he is obliged to spend the bulk of his time and energy and his organization's money in countries which exhibit long-term financial and political stability as well as a sympathetic attitude towards foreign investment. By definition, countries of this type will tend to be those where the greatest exploration activity has been concentrated in the past. In these territories the prospector has therefore been forced into searching for deposits at depth with no surface expression other than possibly, but not necessarily, an encouraging geological environment.

The rate of discovery of significant base metal ore bodies has slowed during the past ten years because of the difficulties in detecting mineralization which is evidenced only by very subtle surface indications or which occurs at depths of several hundred feet or more with no outward indication of its presence other than an ambiguous geophysical or geochemical response. The need to develop techniques for detection at depth has proved a great spur to geologists, physicists, and chemists to improve their concepts and systems to meet this challenge. As a result of this pressure the research worker in exploration geology takes a far more constructive view of the economic significance of his work than in the past. The geological models which are developed have immediate application in the selection of areas for further search and the ideas which are employed are under constant review, both in universities and within the mineral industry itself.

2. Exploration methods

The technological developments mentioned earlier have produced a number of geophysical systems for both airborne and ground use. If it is decided that a project is best tackled by the use of geophysics the exploration geologist is required, probably with the help of a geophysicist, to choose from the numerous methods which are available the one which best suits his technical requirements, the prevailing environmental conditions, and last, but not least, the money he has been allowed for the project (Brant 1967). The balance between the money available and the degree of technical perfection he seeks is one of the most difficult judgements the geologist has to make.

Airborne techniques are judged by their relative depth of penetration, their sensitivity, discrimination, and resolution, and the extent and speed of their ground coverage. Table 6.1 lists airborne geophysical systems which were in use in Canada in 1971 and shows the wide variation in their operating techniques, characteristics, and performance.

An especially important feature of any system is the help it gives the observer in discriminating conductors due to sulphide minerali-

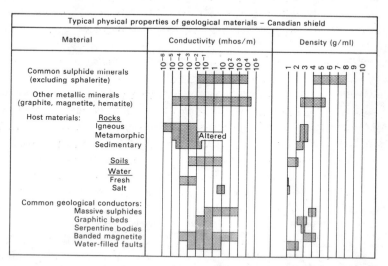

Fig. 6.1. Typical physical properties of geological materials in the Canadian Shield (Paterson 1971).

TABLE 6.1

Airborne electromagnetic systems used in Canada in 1971
(after Paterson 1971)

Category	System	Coil Configuration	Normal Height (ft) Transmitter	Normal Height (ft) Receiver	Freq. (hz)	Penetration (ft)
Rigid-coupled. in-phase/quad. (helicopter-towed)	Lockwood LHEM-200/300	Vertical, coaxial	100	100	4000/ 500, 1000, 2000	300
Rigid-coupled, in-phase/quad. (helicopter-towed)	Scintrex HEM-701	Vertical, coaxial	100	100	1600	300
Rigid-coupled, in-phase/quad. (helicopter-towed)	Keevil/ Barringer	System A: vertical, coaxial.	100	100	918	430
(System B quad. only)	DIGHEM	System B: transmitter vertical, receivers horizontal and vertical (minimum-coupled)				235 Horizontal coil only
(System B quad. only)	Sander	Vertical, coaxial	100	100	1000	260
Rigid-coupled, in-phase/quad. (helicopter-mounted)	TGS/Varian	Vertical, coaxial	150	150	400	365
Rigid-coupled, in-phase/quad. (fixed-wing mounted)	Can. Aero/ Geoterrex/ Scinrex/ Rio-Mullard	Vertical, coplanar	150	150	320	410
Rigid-coupled, in-phase/quad. (fixed-wing mounted)	Can. Aero Canso	Vertical, coaxial	175	175	390	500
Towed-bird, quadrature	McPhar F-400	Transmitter vertical, receiver horizontal	275	125	340/1070	1050
Towed-bird, quadrature	Lockwood LEM-210/ Hunting	Transmitter horizontal, receiver vertical	450	230	400, 2300	1000
Towed-bird, quadrature	Sherritt/ Newmont	Transmitter horizontal, receiver vertical	425	245	385, 1700	1050

TABLE 6.1—*cont.*

Towed-bird, time domain	Barringer INPUT	Transmitter horizontal, receivers horizontal and two vertical. Records total sec. field strength	400	160	300–1900 μ s.	900
Towed-bird, differential	INCO	Records difference between fields in two receivers, excited by separate transmitter coils at different frequencies. First pair vertical coplanar; second pair tiled coaxial	450	175	400–2300 (approx.)	1350
Two-plane, differential	ABEM	Records difference between fields in two receiver coils (one horizontal, one vertical), excited by separate coplanar transmitter coils, operating in quad, phase to one another	262	262	880	1500
AFMAG (dip angle)	McPhar AF-4	Two orthogonal receiver coils, with axes in line of flight	100– 1000	100– 1000	140, 500	large
VLF (dip angle)	McPhar KEM	Two orthogonal receiver coils, with axes in line of flight	100– 1000	100– 1000	15–25k	large
VLF (In-phase/ quad.)	Geonics EM-18	Horizontal receiver, vertical reference	100– 1000	100– 1000	15–25k	large
VLF (In-phase/ quad.)	Barringer RADIO-PHASE	Horizontal and two vertical receivers, vertical electric field reference	100– 1000	100– 1000	15–25k	large

TABLE 6.2

Some ground geophysical techniques available to the exploration geologist

	Penetration below surface	Relative speed	Best target	Strengths	Limitations	Confusing factors
Vertical loop electromagnetics	200 ft	Fast; 2–3 line miles/day	Tabular; steeply dipping conductive bodies	Rapid; good penetration; simple results; minimal noise problems	Non-diagnostic information; insensitive to some orientations of conductors	Formational conductors, barren sulphides
Horizontal loop electromagnetics	100–150 ft	Moderate; 2 line miles/day	Steeply and shallow dipping conductive bodies	Diagnostic information on conductivity width and dip; sensitive to various orientations and lengths of conductors	Limited penetration; slower; noise from overburden conductivity	Overburden conductivity; formational conductors; barren sulphides
Turam electromagnetics	300 ft	Slow: 1–2 line miles/day	Tabular; steeply dipping conductive bodies	Deep penetration, diagnostic information on conductivity and geometry	Slow; expensive; insensitive to some conductor orientations; cumbersome operation	Formational conductors, barren sulphides
VLF electromagnetics	100–200 ft	Fast; 2–4 line miles/day	Tabular; steeply dipping conductive bodies	Rapid; useful for structural information; one man operation	Multitude of structural and topographic responses; limited penetration	Overburden conductivity; topographic features

Method						
Induced polarization	Up to 500–1000 ft for large zones	Slow; 1 line mile/day	Disseminated sulphides	Only method to detect disseminated sulphides and will detect most massive zones as well; deep detection of large zones; description of near surface geology more comprehensive than any other geophysical tool	Slow; expensive; produces many anomalies	Magnetite, chromite serpentine, clay zones, some alteration minerals
Ground magnetics	Variable	Fast; 2–4 line miles/day	Magnetic zones	Rapid; useful adjunct to electromagnetics, particularly in search for magnetic sulphide zones; useful for geological interpretation	Reflects many magnetic rock types other than magnetic sulphides; many deposits of massive sulphide are non-magnetic	Great variety of magnetic geological features
Gravity	Variable	Slow; 1–1¼ line miles/day	Massive sulphide zones	Useful adjunct to electromagnetics and magnetics in search for massive sulphides	Slow; expensive; sensitive to irregularities of bedrock topography and dense geological units other than sulphides; difficult to interpret at depth	Irregular bedrock topography; dense rock units

TABLE 6.3

Geochemical techniques available to the exploration geologist

	Penetration below surface	Relative speed	Best target	Strengths	Limitations	Confusing factors
Stream sediment sampling	Negligible	Fast coverage of areas	Near-surface or exposed oxidizing sulphide zone	Low unit cost; rapid; specific indicator; can detect disseminated or massive sulphides; relatively unskilled labour can be used for collection	Requires well dissected area and good dispersion to assure effectiveness in areas of thick transported cover or cap rock; summer work only in northern regions	Small and/or low-grade sulphide zones; lithological anomalies; 'cultural' anomalies; scavenging of metals by Fe and Mn oxides; metal concentrations in swamps and at precipitation barriers
Soil sampling	Negligible	Moderate coverage of areas	Sulphide zone under residual soil or thin cover of transported soil	Specific indicator; can detect disseminated or massive sulphides; relatively unskilled labour can be used for collection; may also give useful geological information	Masked by transported cover or cap rock; summer work only in northern regions	Small and/or low-grade sulphide zones; lithological anomalies; cultural anomalies; scavenging of metals by Fe and Mn oxides; metal concentrations in swamps and at precipitation barriers

Method					
Rock sampling	Negligible	Slow coverage of areas	Weathered sulphide zones; gossans	Specific indicator of presence of sulphides; geological information	Good exposure of outcrops needed if used as general survey method. Moderately skilled men required for work.	No discrimination in size or tenor of sulphide zones; Fe and Mn scavenging of metals: sample preparation expensive. Sometimes misleading owing to complete leaching of base metals.
Water sampling					
Vegetation sampling	Normally used only in specialized surveys to deal with very specific problems.				

zation from anomalies which reflect other conductive situations not necessarily associated with sulphides. Fig. 6.1 shows the overlap in physical properties which can confuse the interpretation of data and demonstrates the importance of some degree of discrimination in the airborne electromagnetic systems. In order to improve the discrimination of these techniques much work is being carried out on the design of the equipment and on the processing of the data collected.

Ground geophysical methods form an important part of the techniques available to the exploration geologist when following up airborne results and defining targets prior to drilling. As with airborne techniques, a wide range of instrumentation is available and the geologist must be familiar with the limitations and advantages of each method and be able to weigh the technical aspects of a method against the costs of employing it. Table 6.2 shows in a very generalized way some of the techniques available and the parameters which govern their use.

Chemistry has also played its part to an increasing degree in the last few years as one of the major tools available to the geologist in the search for minerals (Booth 1971). Applied geochemistry has benefited greatly from the application of rapid analytical techniques employing atomic absorption analysis as well as an increasing understanding of the mechanisms of metal dispersion in stream waters and sediments, and in soils and rocks. Here again the geologist must form an opinion as to the applicability of the techniques he has at hand in terms of the metal sought, the environment he is working in, and the amount of money available.

Table 6.3 summarizes the major geochemical techniques available to the exploration geologist. To this list can be added the techniques of vegetation and water sampling, various methods of lake-silt sampling, and air and soil-vapour sampling. The last-named techniques are still largely in the development stage, but with improved understanding of the mechanisms by which trace-metal dispersion takes place, some of them can soon be expected to become useful to the prospector.

As with geophysics, the various geochemical techniques are continuously under review and much work has been carried out on the processing of data using computers and advanced statistical methods. It should be added here that with all interpretative processes the end-product is only as good as the quality of the input

TABLE 6.4

Schematic representation of a typical base metal
exploration programme using a predominantly geophysical approach

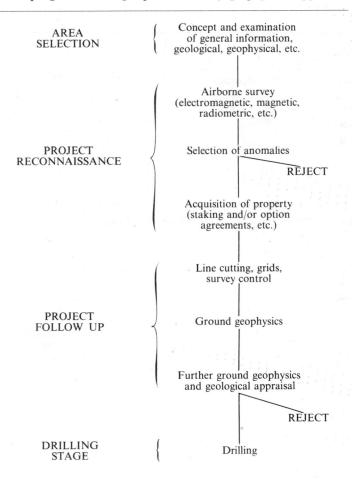

TABLE 6.5

Schematic representation of a typical base metal exploration programme using a predominantly geochemical approach

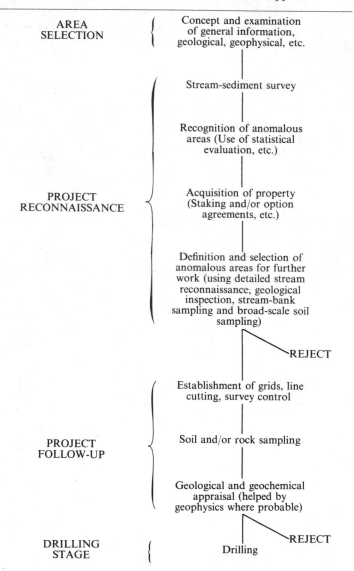

AREA
SELECTION
{ Concept and examination
of general information,
geological, geophysical, etc.

PROJECT
RECONNAISSANCE

Stream-sediment survey

Recognition of anomalous
areas (Use of statistical
evaluation, etc.)

Acquisition of property
(Staking and/or option
agreements, etc.)

Definition and selection of
anomalous areas for further
work (using detailed stream
reconnaissance, geological
inspection, stream-bank
sampling and broad-scale soil
sampling)

REJECT

PROJECT
FOLLOW-UP

Establishment of grids, line
cutting, survey control

Soil and/or rock sampling

Geological and geochemical
appraisal (helped by
geophysics where probable)

REJECT

DRILLING
STAGE
{ Drilling

information and data-processing systems are merely a tool to help the operator reach a conclusion.

3. The Exploratory programme

For the sake of simplicity the typical base metal exploration programme is here divided into four stages. Examples of programmes which rely predominantly on either a geophysical or a geochemical approach are set out in Tables 6.4 and 6.5. In practice, short cuts are taken within these formats when the geologist sees that it is possible to make them without jeopardizing the end-result. As indicated above, the direction given to the programme during its course will depend upon the experience of the man handling the work—and at times upon his intuition.

3.1. Area selection

It is the area selection stage which essentially sets the boundaries of a general area within which the chances of finding an economically viable mining operation are judged to be higher than average. The geological work normally proceeds on the understanding that the exploration company in question has already reached a favourable assessment of the future political stability of the area concerned.

The ground selected during this stage of the work can be anything from one hundred to several thousand square miles in area. The choice depends on numerous factors, among which the most important would be the quality of information available on the area, its accessibility, its geological complexity, and the local laws governing land tenure.

In the early stages of a grass-roots exploration study (i.e. one where the company is essentially working from first principles) the work is normally left to one or two individuals within the exploration group. These men will have a particular talent for area selection based on a sound knowledge of geology and the factors governing the formation of an ore deposit. They also have to be well versed in their company's requirements and operations, and they represent some of the exploration group's most valuable personnel.

Much of the area selector's time is spent in the library sifting government and private reports on the general area in which the company wishes to concentrate. In some countries where the mineral industry is well developed, as in Canada, information on regional geology and government-sponsored geophysical data are available in some detail. In these cases the basic information the area selector requires is comprehensive and easily obtained. The company is not forced to seek the bulk of the data itself by making photographic reconnaissances, extensive field trips, and preliminary surveys.

Most government survey departments are less well established than this and a company may well have to make the preliminary surveys itself, guided in the first instance by very scanty information and theory. These preliminary surveys are often controlled by aerial photographs to help in the correlation of existing geological and geophysical information and data obtained from brief field visits to the areas in question. In these circumstances the company, in addition to building up a generalized geological map, may decide to fly magnetometer and other geophysical surveys to help in the interpretation of the geology. Each region of interest has to be checked in this way before the final selection can be made.

At times even the better surveys are not correct in detail, since they are generally carried out on a very broad scale. Thus, rocks can be identified wrongly on the maps and assumptions made by the original surveyor can turn out to be incorrect when a company moves into the area on its project follow-up stage. It is therefore possible for the area selector to be misled by his data. He will normally try to reduce the chances of this happening by making several short field trips to areas in which he is interested to check on the information he is using.

The area selection stage of the exploration programme is all-important if one intends to place oneself in the area where conditions are most favourable to the development of a mine. Of all the various facets of an exploration programme it is this preliminary stage, especially in a competitive situation, which has the most important influence on attaining ultimate success. The geological team which can formulate a new approach in the initial selection stage or can pick out a new and hitherto unnoticed area of high promise has a commanding advantage over its rivals.

The competition for good ground in the industry is very strong. This is due in part to the uniformity of the approach adopted by the mining companies in the application of area selection methods and exploration techniques. This uniformity in thinking often tends to lead them to the same conclusion at the same time and also allows them to assess any advances in another company's thinking very rapidly. Companies tend, therefore, to play follow-the-leader in the search for new mines (Mackenzie 1972). Particular theories, areas, and techniques often become fashionable throughout the whole industry at the same time, once again pushing companies towards a certain line of geological thought and invariably to the same areas of the earth's crust. Fashions come and go, however, and the area selector must therefore be very level-headed in the use he makes of the various criteria at hand when drawing up his recommendations.

Currently, for example, he would consider the presence of the relevant metallogenic province, the general age of the rocks, the presence of areas of deformation and faulting, and the presence of an assemblage of rocks which is similar to known deposits of the type he is seeking. In addition, since plate tectonics are in vogue at the present time, some application might also be made of these ideas. Similarly the control exerted by systems of intersecting fracture-patterns on the deposits of mineralization are given credence by some geologists. In general it seems that the truth is touched on by most of these theories but in no case can the location of ore in the earth's crust be wholly and satisfactorily explained by any one of them (Booth 1971).

To summarize, therefore, area selection can entail office studies of geological information on a broad scale and geological compilation from government maps and reports as well as other sources. In the absence of other data, area selection will require reconnaissance, geological mapping, and prospecting, possibly backed up by broad-scale airborne geophysics and stream and soil geochemistry.

3.2. Project reconnaissance

Having outlined the general area of interest, which may be up to several thousand square miles in extent or in some cases even more, the company will embark on the next phase of the programme, which is project reconnaissance. The object of this phase is to establish, within the boundaries set by the area selectors, the project areas which are most likely to reward the later intensive and expensive phases of the work.

In Canada, especially in the Pre-Cambrian Shield, the initial stage of the project reconnaissance involves the use of airborne geophysical instruments. These instruments have been developed to provide accurate and relatively inexpensive information regarding the presence or absence of anomalously conductive and magnetic zones in the earth's crust. They are useful for checking areas of interest where access and geological expression are poor, such as in the swamps of northern Canada, and where a ground examination would be very difficult, time-consuming, and relatively expensive. Surveys are flown in areas which show particular geological promise and may involve several hundred square miles of survey at one or more locations out of the two or three thousand square miles considered in the original area selection studies. As indicated elsewhere, airborne electromagnetic instruments do not have universal application however, owing to problems associated with high surface conductivity, deep surface weathering, rugged topography, and so on. Where these problems exist, a combination of geological and geochemical methods on a broad scale might be used over selected areas to the same effect as the airborne geophysical survey.

After this initial survey stage has been completed, if the preliminary information is promising, a project geologist is appointed. He would normally have had several years' field experience and would probably have worked in similar terrain before. At the start of the reconnaissance work the project geologist works closely with the area selector but, in the later stages of project definition and drilling, the project geologist takes on the entire responsibility for the field-work on a programme.

The project geologist on a job such as the one we are discussing will have, according to the terrain in which he is working, one or two mapping crews of two men each. One man in each team will be a geologist. If the project is geophysically orientated the project geologist might also have two teams of geophysical instrument operators, a team of line cutters who will mark out the ground for the various surveys, a cook, and possibly one or two general hands.

Table 6.6 shows a typical organization chart for a relatively large grassroots base-metal exploration project in Canada with an indication of the type of equipment for which the geologist in charge will be responsible. The format of this organization will change if the crew is working in an arid climate or in mountainous terrain and will also depend on the emphasis to be placed on geology or geochemistry.

In all, the project geologist may have charge of ten to fifteen men

TABLE 6.6
*Generalized organization on a grassroot base-metal
exploration project in Canada*

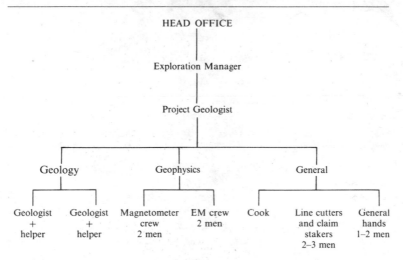

Sundry equipment

Accommodation: Tents, trailers, huts.
Boats: Canvas, rubber, aluminium, etc.
Fixed-wing aeroplanes: Usually chartered; landing on water in summer and ice
  in winter.
Helicopters: Year-round usage on charter: more expensive than fixed-wing aero-
  planes.
Radio: Two-way voice radio.

and in addition to his primary role his job calls for a high degree of
practicality in everything from machinery repairs to cooking. He
has to manage and coordinate the activities of the various crews
and keep up a steady commentary on the results achieved through
letters, reports, and maps to his head office. Most of the more
remote camps are in touch with their offices by radio, and aeroplanes and helicopters play an important role in servicing exploration camps in all parts of the world.

Accommodation is adapted to the country climate. In Canada,
much of the initial work is done from large tents, sometimes set on
plywood floors (Fig. 6.2). These are provided with efficient oil or
wood heaters in winter, for in many areas temperatures drop to
$-20°$ to $-40°F$ at night. In hot and arid climates accommodation

127

Fig. 6.2. Canadian Bush camp in winter. (Photograph: Robert C. Ragsdale, Toronto.)

is sometimes in tents but portable huts and trailers are becoming increasingly common.

The project reconnaissance stage is very important to the success of an exploration programme since it is at this stage that the geologist picks, for example, the most economically significant 10 square miles of ground in the original 100 square miles selected on the basis of general geological information in the area selection stage of the work. If only one potential mine exists in the 100-square-mile area, the searcher makes the assumption essentially that the 10-square-mile area picked is the one in which the mine is most likely to exist. It can be seen therefore that the project reconnaissance must be carried out with great care. It is also true to say that companies will reassess their original area selection information from time to time in the light of the knowledge gained later in the programme and in this way may select additional project areas. Often, however, second runs at the reconnaissance stage are not possible; competitors may have taken up the alternative ground or money may not be available for a second try.

In Canada the area selection and project reconnaissance stages are usually closely allied. The whole process from the beginning of area selection, the subsequent airborne geophysical surveys, the interpretation of the data, and the staking of claims in the areas of interest commonly takes only two to three months.

In Australia and arid areas of South Africa, exposure of diagnostic rocks is generally better than in the Canadian Shield and this, combined with the fact that the high surface conductivity of the soils in these dry climates modifies the effectiveness of electromagnetic systems, tends to place the onus of the reconnaissance work on detailed geological mapping and possibly soil and stream geochemistry. Airborne magnetometers have proved to be very useful reconnaissance tools in these conditions. A low-level technique using a sensing head, which is towed some 125 ft above the ground on flight lines only a few hundred feet apart, has been especially effective. The resulting data from surveys of this type are very detailed and are approximately comparable to those obtained at greater expense on the ground.

In the open expanses of desert areas, aerial photographs have become widely used, both in black and white and colour; the latter add a new dimension to geological interpretation of rock-types and structure.

Fig. 6.3. Horizontal loop electromagnetic operator in northern Canada. (Photograph: Robert C. Ragsdale, Toronto.)

3.3. Project follow-up

During the follow-up stage of the work the project geologist is concerned with defining the nature of the targets outlined during project reconnaissance. In the Canadian Shield, where programmes are normally geophysically based, follow-up crews operate ground magnetometer, electromagnetic (Fig. 6.3), and gravity instruments in an effort to track down and define the airborne geophysical anomalies. Geological mapping will also be used in this stage of the work if outcrops exist; this will greatly improve the interpretation of the geophysical results.

In Australia and South Africa, where the terrain tends to be more open, geological work combined with soil and rock geochemical prospecting and some ground geophysics are used to develop the areas picked in the reconnaissance stage (Cowan *et al.* 1974). Also in common use are trench diggers, soil augers, and light percussion drills. These mechanical aids enable the geologist to obtain samples below the zone of intense surface weathering which tends to obscure the significance of outcrops in arid climates.

The follow-up stage must be carried out painstakingly and must to some extent be geared to a target of predefined dimensions and tonnage. The size and type of target must be defined in general terms for the exploration geologist by his head office. This will allow him to design a follow-up programme in a way which will help him to interpret geological, geophysical, or geochemical evidence in order to discriminate between deposits of economic size and importance on the one hand, and deposits with little mineralization or of the wrong type on the other. The 'net' cast by the geologist may, of course, detect a small-ore body by chance, which should be examined with a view to developing it, but the overall exploration programme should be designed to minimize chances of missing an ore-body of the required size. The concept of 'mesh size' in this exploration 'net' is very important and is also considered carefully during the project reconnaissance stages of the work.

The time spent in follow up can vary greatly depending on the initial complexity of the geology, the quantity and quality of the data obtained in the reconnaissance phase of the work, the size of the area, the climatic conditions, and the communications. On average it normally takes one to three years to follow-up adequately and assess all the responses detected during the geological and geochemical reconnaissance of an area of 100–200 square miles.

3.4. The drilling stage

The last and most expensive stage of the exploration programme, at least on a unit basis, is the drilling stage. Normally at least a quarter of the total exploration budget in a grassroots programme of the type discussed here is spent on this work, and allowance must be made for these activities when drawing up the programme and estimates for an area.

Drilling represents the only completely satisfactory way of testing for mineralization in depth other than by actually mining. The drilling methods available to the exploration geologist are broadly of three types.

Auger-type drills will sample soils and a limited depth of weathered rock. These are light drills, often mounted on a tractor or on a small easily moved trailer. They form an important part of the equipment used by the geochemist in arid open terrain.

Percussion and rotary drills are produced in several sizes. Percussion drills are airflushed; rotary drills may use a water or mud lubricant. The sample is produced as a small chip of rock and there are problems in locating the exact position from which it was derived. Percussion drilling can be seriously limited by the presence of water in the hole, and thus these drills are not normally employed if the hole is to penetrate very far below the existing water-table. Rotary and percussion holes, however, have the advantage over diamond drilling methods in that they are very rapid and relatively cheap to operate and they can provide a hole of large diameter at much lower cost than a diamond drill. In practice the larger percussion drills are often used early in drill programmes to take a first look at the mineralization at depth. If these results are encouraging the geologist calls for a diamond drill. Because of the heavy compressors and trailers required for operating percussion drills they tend to be used only in open country where access is relatively easy.

Diamond drilling is the most common method of acquiring information about mineralization at depth (Fig. 6.4). Here the drill cuts a continuous cylindrical section out of the rock and thus provides the geologist with a sample which can be accurately located in the hole and can supply structural, petrological, and assay data on the section under examination. Diamond drilling technology has advanced rapidly in the past ten years. However, it still lags behind the sophistication of the oil drilling industry and its massive rotary rigs. Diamond drilling is a slow procedure and progress ranges between 30 and 80 ft in 24 hours, depending on ground conditions and the depth at which the drill is operating. The work is often

Fig. 6.4. Diamond drilling rig set up on an inclined borehole exploring for nickel in Western Australia. (Photograph: Australian Selection Pty. Ltd.)

very arduous for the men who carry it out and they have to compete with great extremes of temperature and weather, often working on exposed sites. Because of the remote situation of many of the drill sites, equipment is both expensive and difficult to maintain and the men themselves are hard to hire or keep. The costs of diamond drilling currently range between £5 and £15 per foot, according to the country in which the work is being carried out and the situation of the property to be tested. Thus a 500-ft hole can cost up to £7500; and normally at least two or three holes are required to test a target adequately in the first instance.

Because of the expense of drilling a hole with rotary, percussion, or diamond drills, every attempt is made in the follow-up stage to reduce the risk of failing to detect mineralization and, one hopes, a new mine. In some ground conditions it is possible to use light-weight diamond drills or percussion and rotary drills, all of which are less expensive than the conventional diamond drill. As in all other phases of the exploration programme, the geologist must balance the quality of the information he requires against the cost of acquiring it.

In the course of the drilling, the project geologist will have to supervise the logging of the core and its storage as well as providing the preliminary interpretation of the results (Fig. 6.5). If the core is mineralized it will be split and half will be sent for assay. The geologist will be responsible for siting subsequent holes and for seeing that they are drilled correctly and efficiently. In the span of a few seasons' work a field geologist acquires a detailed knowledge of drills, drill techniques, and the administration of drilling programmes.

4. Conclusion

The foregoing account of the responsibilities of a field geologist in a base metal exploration programme is, as has already been stressed, rather superficial but it should give an idea of the great variety of activities which this branch of the geological profession provides. The examples quoted from Canada and the more arid regions of Australia and South Africa represent extremes of climate and operating difficulty. More temperate environments have their particular problems, both technical and physical.

Grassroots exploration has infinite variety and taxes to the full the resourcefulness of the men involved. Few who become immersed in it ever fully throw off its fascination.

Fig. 6.5. Logging core in Western Australia. (Photograph: Australian Selection Pty. Ltd.)

Mineral exploration

Acknowledgement

The author acknowledges with thanks the basic information provided by the Selection Trust group of companies and specifically the Selco Mining Corporation.

References

BOOTH, J. K. B. (1971). The influence of the natural abundance of metals on exploration. *Can. Min. metall. Bull.* **64**, 128–33.

BRANT, A. A. (1967). Some considerations relevant to geophysical exploration. *Can. Min. metall. Bull.* **60**, 54–62.

COWAN, D. R., ALLCHURCH, P. D., and OMNES, G. (1975). An integrated geoelectrical survey on the Nangaroo copper–zinc prospect, near Leonora, Western Australia. *Geoexploration* **13**, 77–98.

MACKENZIE, B. W. (1972). Corporate exploration strategies. *Application of computer methods in the mineral industry. Proc. 10th. Int. Symp.* Johannesburg.

PATERSON, N. R. (1971). Airborne electromagnetic methods as applied to the search for sulphide deposits. *Can. Min. metall. Bull.* **64**, 29–38.

7. The metallogeny of Britain

1. Introduction

Metals occur naturally in the form of minerals which may be native elements, or oxides, sulphides, tellurides, arsenides. carbonates, sulphates, or hydroxides. An ore is a mineral or aggregate of minerals which can be worked for the recovery of a metal or metals with profit or hope of profit. The British Isles have been considerable producers of tin, copper, zinc, lead, silver, and iron, dominating world production of each of these metals except zinc and silver for some part of the nineteenth century, though with outputs that look small in comparison with modern standards. Workings are widely scattered over the upland regions from Cornwall to the Scottish Highlands and a great deal is known about the deposits which were worked. Copper is no longer being produced, but resources of this metal are known to exist; the remaining metals are all being mined, some of them, however, on a small scale only. One of the questions of the moment is: are there worth-while quantities still to be found? As the economic situation changes, deposits are once again being actively sought, with financial encouragement from the government.

It may safely be assumed that in the course of a thousand years of mining (perhaps three thousand years in the case of tin) all the easily found ores in Britain have been discovered. If the concealed deposits are now to be found, intelligent use must be made of the geological information provided by past prospects and workings. In this chapter we attempt a broad survey of some of the principal factors that have controlled the mineralization of Britain. One consideration is basic to the inquiry: ore deposits in Britain, as elsewhere, are very small features of the lithosphere, but they represent concentrations of the useful and associated elements 5 to 2000 times greater than the average content of these elements in the crust. The problem that faces the metalliferous geologist is to explain and understand the numerous and complex processes that lead to ore-formation. Metallogeny demands a series of remarkable coincidences of processes, not evidenced in average rocks.

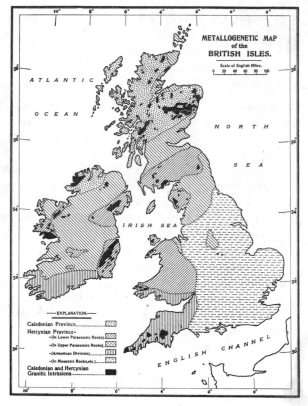

Fig. 7.1. Metallogenetic map of the British Isles by A. M. Finlayson. (Reproduced, by permission, from *Q. Jl geol. Soc. Lond.* **66**. 1910.)

2. Mineral provinces

The study of the regional distribution of ore deposits in relation to major geological features, initiated by Louis de Launay (1905–13) in France, was first undertaken in this country by A. M. Finlayson (1910*a*). His map is reproduced as Fig. 7.1; and for comparison Fig. 7.2 shows the principal mineral fields. De Launay was much

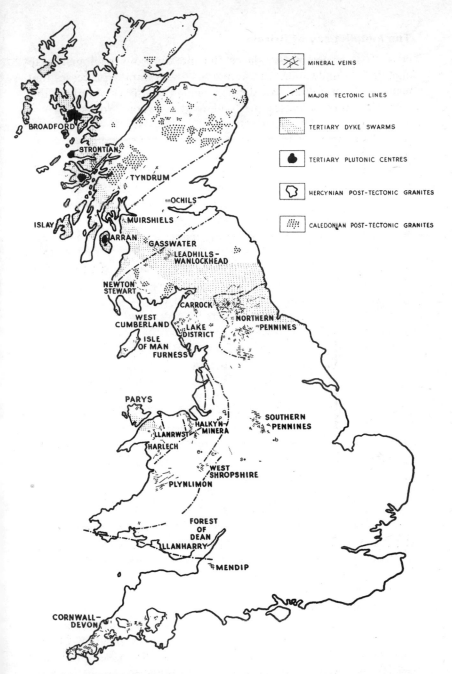

Fig. 7.2. Distribution of British epigenetic ore deposits by K. C. Dunham. (Reproduced, by permission, from *Trans. Geol. Soc. Glasgow*, **21**, 1952.)

influenced by the fundamental structural work of Edward Suess (1909) to whom we owe the establishment of the major orogenic belts of the continents. In Britain there are parts of two of these: the Caledonian, trending NE–SW, affecting Pre-Cambrian and Lower Palaeozoic rocks in Scotland and continuing through the Lake District and most of Wales; and the Hercynian, trending E–W and affecting Devon, Cornwall, and the southernmost part of Wales. The movements which made the Caledonian Mountains are considered to have culminated in late silurian–early Devonian times, and those forming the Hercynian ranges in late Carboniferous–early Permian. The concept that underlies the idea of mineral provinces is that the concentration of the elements is the result of special physical and chemical activity associated with the orogenies. Earth stresses, leading to the formation of fracture patterns under tension, and therefore permitting the ingress of mineralizing fluids, are the chief physical controls, while igneous magmas undergoing movement, intrusion, extrusion, and crystallization provide the chemical and thermal sources.

The oldest Pre-Cambrian rocks in Scotland, exposed in the Outer Hebrides and the north-west Highlands record, as we now know, the effects of three major tectonic episodes earlier than the Caledonian, but apart from a few pegmatites, one of which has been worked for ceramic feldspar, mineralization is curiously lacking here. Finlayson could identify only one Pre-Cambrian ore deposit, the tin–magnetite deposit of Carn Chuinneag. Nowadays we regard the gneissose granite in which this deposit occurs as one of the 'older' granites, intruded *pari passu* with the Caledonian movements, and the mineralization is thus regarded as belonging to the Caledonian province.

Finlayson's map (Fig. 7.1) shows the Scottish Highlands and Islands, together with north-west Ireland as a Caledonian mineral province. He mentioned particularly the molybdenite scattered through some of the granites, the chromite in Unst, the nickeliferous pyrrhotite of Loch Fyne, and the sulphide impregnations ('fahlbands' in the Norwegian sense) around Loch Tay and in Cowal. No substantial ore deposit has yet been found among these, but some of them are of types similar to large deposits worked in the continuation of the Caledonian province through Scandinavia. Prospecting in the concealed areas is now recognized to be desirable. The Avoca pyritic copper deposit, in Wicklow, properly belongs here. The alluvial gold, for example around Helmsdale, must have been derived from Caledonian sources.

Those parts of the former Caledonian mountains now forming the Southern Uplands, the Lake District, and much of Wales, where despite intense isoclinal folding, the metamorphic grade is low compared with the Highlands, contain far more known ore deposits. Two can be proved to be of pre-Carboniferous age. The tungsten veins of the Grainsgill, near Carrock Fell, Cumbria are so obviously linked with the Skiddaw granite through marginal greisen as to leave no doubt as to their age; and the Parys Mountain copper deposits in Anglesey (which for a time in the mid-nineteenth century controlled the world price of the metal) are of proved Lower Palaeozoic age since pebbles derived from them occur in near-by basal Carboniferous conglomerates. Finlayson did not, however, accept the other numerous lead–zinc and copper districts situated in the Lower Palaeozoic rocks (see Fig. 7.2 and Table 7.1) as of Caledonian age, but instead assigned the metallization of these to the next later orogenic epoch, the Hercynian. His reason was that the vein-fissures are known to be 'post-Carboniferous or pre-Triassic' in the older rocks of the Lake District and of North Wales, while in the Isle of Man the fissures are later than the pre-Carboniferous disturbances. He was impressed with O. T. Jones's demonstration (subsequently published, 1922) that the vein fractures in Cardigan cut across and shift the Caledonian folds, and held this to be true in Merioneth and Shropshire also. Further, he pointed out that adjacent to the Flintshire and Minera lead districts in Carboniferous Limestone, similar veins could be seen in Wenlock Shales, and he equated the barytic lead veins of Greenside and Thornthwaite in the Lake District with those in the Carboniferous rocks of the Pennines on Dufton Fell and Alston Moor. For these primarily structural reasons, Finlayson's map (Fig. 7.1) shows what are, from the point of view of sedimentation and folding, Caledonian areas as belonging to the Hercynian mineral province. He may also have been influenced by the obvious lack of any consistent spatial association between the Caledonian post-tectonic granites and the mineral fields, save in the single case of Carrock, and the feeble mineralization at Shap. Of the 45 stocks and small batholiths exposed in the British Caledonides all but a very few are totally lacking in evidence of associated hydrothermal metallization. It is tempting to conclude that these granites were, in

TABLE 7.1

Characteristics of British orefields

Orefield	Ore elements	Vein directions	Country-rock	Bearing horizons	Deepest mine (m)
Cornwall–Devon	Su, W, Cu, Zn, Pb, As, (Sb, Fe, F, U, Ba)	ENE WNW E–W (NNW)	Palaeozoic slates, diabase, granite		1006
N. Devon–W. Somerset	Fe, Cu, Pb	WNW	Devonian slates sandstones		254
Mendip	Pb	E–W (N–S)	Carboniferous Limestone (K–D zones) (Also Mesozoics)		146
Glamorgan	Fe		Carboniferous Limestone		290
Forest of Dean	Fe	—	Carboniferous Limestone	Crease Limestone	278
Plynlimon	Pb, Zn (Ba)	ENE E–W	Silurian sedts Ordovician	Frongoch formn Van formn	305
West Shropshire	Pb, Zn, Ba	E–W	Ordovician sedts and volcanics Longhope	Mytton Grit	503
Berwyn Hills	Pb, Zn	E–W	Llandeilo slate, rhyolite		100
Harlech	Au, Cu	NE E–W	Cambrian sedts	Clogau, Maentwrog formns	155
Llanwrst	Pb, Zn	N–S E–W	Bala sedts and volcanics		174
Anglesey (Parys)	Cu (Pb, Zn)	NE	'Felsite' in Lower Palaeozoic sedts		320
Halkyn–Minera	Pb, Zn	ENE WNW (NNW)	Carboniferous Limestone	'Intermediate' (D?) Lst	366

Locality	Metals	Trend	Host rocks	Horizon	
Cheshire Basin	Cu, V	—	Trias sandstone	Sandstone	305
Derbyshire Pennines	Pb, Zn, F, Ba (Cu)	ENE WNW	Carboniferous Limestone	Top Lst and up to 4 others	300
Northern Pennines	Pb, Zn, F, Ba (SO_4, CO_3) (Cu)	ENE WNW (NNW')	Carboniferous Yoredale facies	6 thin Lsts / 5 Ssts / 1 Dolerite	730
Lake District	Cu	E-W	Lower Palaeozoic slates and volcanics	Various	613 553
Isle of Man	Pb, Zn (Cu)	E-W N-S	Slate and granite		198
West and south Cumbria	Fe (Cu)	NNW EW	Carboniferous Limestone	Red Hill Oolite (C, S)	438
Leadhills–Wanlockhead	Pb, Zn	NNW NNE	Caradoc below thrust	Greywackes	109
Newton Stewart	Ni, Pb	WNW	Diorite / Dolerite dyke in Silurian	Dyke	
Midland Valley					
Gasswater	Ba Ba	WNW	Old Red Sst / Lr. Carb volcs	Sandstones	
Ochils	Pb, Ag	NW E-W	ORS volcs		100
Hilderstone	Ag, Ni	E-W	Dolerite dyke	Dyke	
Arran	Ba	N-S NW	Old Red Sst	Sandstones	
Highlands and Islands					
Islay	Pb	Various	Dalradian Lst	Limestone	274
Tyndrum	Pb, Zn	NNE	do., Qtzt, Schist		
Strontian	Pb (Ba, Sr)	E-W ENE	following camptonite dyke in Moine gneiss	Dyke	

fact, dry. Even in the Isle of Man, where one mine (Foxdale) was worked in Caledonian granite, the granite is not regarded as the source.

Quite a different picture is presented by the Cornubian peninsula, where a series of granite stocks from Dartmoor to Land's End have invaded tightly folded Devonian and older slates ('Killas'). This is the Armorican division of Finlayson's Hercynian province (Fig. 7.1) and it is, of course, Britain's most highly productive metal field. A recent estimate (Morgan 1974) showed that of a total British production to date in terms of 1973 prices of non-ferrous ores worth £5·95 billion (10^9), £4·34 billion came from south-west England. These figures are of course affected by the current high price of tin, but they serve to emphasize the importance of the Armorican subprovince. Here it is easy to see that the numerous veins are related in some way to the underlying batholith. The presence of abundant tourmaline, fluorite, topaz and kaolinite point to a rich concentration of 'volatile fluxes'—boron, fluorine, chlorine, carbon dioxide, and water—during the closing stages of crystallization of the granites. The concentration of the metals was primarily effected, in the view of most observers, during these stages. The whole district is one of the classic cases of 'wet' granites yielding a long suite of igneous–hydrothermal minerals. Finlayson associated with this southern subprovince, the outlying lead–zinc ores in Carboniferous Limestone of the Mendips, the Exmoor mineralization, and the copper, zinc, and lead veins of Kerry, Cork, and Waterford in Ireland. The third subprovince of Finlayson's Hercynian embraces the most important lead–zinc districts of the U.K. and I think he would have extended it to cover the great new discoveries of lead, zinc, copper, and barytes made since 1960 in Ireland. The country-rock in all these is Carboniferous Limestone. The individual fields are: the Northern Pennines, where thin limestones and some sandstones in the cyclothemic Yoredale facies (Viséan and Namurian) are the host rocks; the Derbyshire Pennines, where the uppermost limestone above the first inter-bedded lava ('toadstone') and below the Edale Shale has so far yielded most of the production; and the Halkyn–Minera district, with the uppermost beds of a thick limestone series mineralized. Extensive replacement deposits accompany the very numerous veins in both parts of the Pennines, forming flats, pipes, and enlarged vein-fissures. It is probably no accident that fluorite is an abundant mineral in the Pennine fields, in sharp distinction with the other limestone districts and with all the fields in Lower Palaeozoic rocks. To this subprovince Finlayson also assigned the hematite iron ore districts of West and South Cumbria and Glamorgan, and the limestone ores of the Forest of Dean.

In a second paper, Finlayson (1910b) considered the problems of the lead–zinc deposits in the Hercynian province situated remote from the exposed granites of Cornwall–Devon. His conclusion is worth quoting:

> The conclusion that all the ores are of deep-seated origin, as well as the depths to which the veins extend, indicate that the waters are in the first place, 'juvenile', being probably hydrothermal after-effects of the deep-seated igneous disturbances and intrusions. This is supported by the occurrence of fluorite.... Further, it is not likely that meteoric waters played a considerable part in the primary ore deposition in the slate.... In limestone areas, however ... a consideration of the facts suggests that underground atmospheric waters, mingling with the 'juvenile' waters have played an important part in ore-deposition.

Rather surprisingly, Finlayson (1910a) included as a fourth subprovince of his Hercynian all the remaining parts of the country: that is those underlain by rocks ranging in age from Permian to Tertiary. The deposits in the Mesozoic cover of eastern Midlands and southern England include the bedded cupriferous sandstone of Alderley Edge and Mottram St. Andrews in Cheshire, Whitchurch, Grinshill near Shrewsbury, and Felton near Oswestry. Otherwise, there are trifling quantities of blende, galena, and other sulphides in Keuper, Lias, and Lower Oolite of several counties. These he took to represent the highest zone of Hercynian metallization, indicating that the process went on well into the Mesozoic era.

We should also note that the Mesozoic cover contains the only important metal deposits that everyone accepts as of sedimentary, syngenetic origin. These are the ironstones of the Jurassic and Cretaceous, including Cleveland (Middle Lias), the Marlstone Rock-bed of Lincoln–Leicester–Rutland and Banbury (Middle Lias), Frodingham (Lower Lias), Northampton Sand (Inferior Oolite), and Claxby (Lower Cretaceous). Some enrichment in non-ferrous metals is known to accompany at least one of these, the Cleveland Ironstone.

3. Ages of mineralization
With the framework provided by Finlayson's pioneer work before

us, we may now proceed to review the significant advances made during the past 65 years. Many problems remain to be solved, here as elsewhere in this difficult field of research.

If a convincing case is to be made out for the existence of metallogenic provinces associated with the major orogenies or, for that matter, with plate margin conditions (as many present authors seek to justify), the establishment of the age of emplacement of the orebodies is of critical importance. We have already seen that Finlayson relied on structural and stratigraphical arguments for this purpose. A fuller survey of the geological evidence, mainly derived from the extensive investigations carried out by the Geological Survey under the stimulus of two world wars (Dunham 1952b) led to the following conclusions: (1) the proved Caledonian deposits are Parys Mountain, Carrock, and Shap, together with a few minor deposits in Scotland associated with the 'newer' granites such as Pass of Ballater and Abergairn; (2) conclusive evidence of the age of the other non-ferrous mineral fields in Lower Palaeozoic rocks is lacking, but the close relationship between the main mineralization epoch in Cornwall and Devon and the Armorican granites is well established. There is evidence that the lead–zinc–fluorine–barium mineralization in the Pennines was still in progress in Zechstein (Upper Permian) times; (3) Mesozoic, chiefly Triassic, sandstones which contain copper, vanadium, and barium deposits are analogous with the 'Red Bed' type of the Colorado Plateau and parts of continental Europe; (4) Mineralization associated with the Tertiary igneous centres or dyke-swarms is minimal. The epigenetic ferric ores in Carboniferous Limestone, remote from Tertiary magmatism, may have been introduced during this epoch, but by downward infiltration of mineralizing solutions from the Permo-Trias into the Carboniferous Limestone.

In passing we may note that prospecting during the past decade has revealed low-grade disseminated sulphide bodies which may some day be workable for copper in a north Wales diorite, and for nickel in basic igneous rocks in north-east Scotland; both are presumably of Caledonian age.

Hope of considerable progress towards precise dating came with the elaboration of methods of isotopic analysis of ores. Stephen Moorbath of Oxford was the pioneer (1962) in the U.K. He determined lead isotope abundances on 98 samples of galena representative of all the British non-ferrous fields, interpreting the results according to the equation for the Holmes–Houtermans model, and using Ivigtut (Greenland) lead as his standard. He has never claimed that the results do more than indicate the date of original separation of the lead from its source. It must also be emphasized that the lead–lead method is not an 'absolute' one like U–Th–Pb, K–Ar or Rb–Sr. It depends upon the validity of the model adopted.

C. J. V. Wheatley (1971) in a thoughtful paper on the mineralization of the southern Caledonides has set out the results on a map. Leadhills–Wanlockhead (310 ± 60 to 320 ± 50 m.y.), all but one result for the Lake District (470 ± 50 m.y.; 340 ± 70 to 210 ± 70 m.y.), Llanwrst (340 ± 70 m.y.), and West Shropshire (280 ± 70 m.y.) can be held to support Finlayson's thesis, but the model ages for Harlech (420 ± 50, 470 ± 50 m.y.) and Plynlimon (360 ± 60 to 450 ± 50 m.y.) do not. For Parys Mountain, the model age of 300 ± 40 m.y. is too low to be consistent with the geological evidence, but Avoca at 430 ± 40 m.y. gives a Caledonian age. The results for the northern Pennines coincide within the limits of experimental cover with those for Devon and Cornwall, but Derbyshire, at 150 ± 80 to 220 ± 70 m.y. is surprisingly different, having regard to the geochemical and other similarities between the Pennine fields. However, as noted by Finlayson and myself (Dunham 1952*b*) there are striking and consistent differences in silver content of galena between northern and southern Pennine galenas.

Unfortunately the lead–lead method of model age determination is very sensitive to the constants employed, and on these grounds Moorbath's interpretation of his results has been seriously challenged, for example by R. H. Mitchell and H. R. Krouse (1971). Using isotopic ratios for primaeval lead based on the Broken Hill (New South Wales) galena, and using decay constants different from those employed by Moorbath, their recalculations of his data give model ages considerably younger than the geological evidence justifies in the case of the limestone districts. Then, they conclude, the lead in these is similar to that in the Mississippi Valley deposits; it has been contaminated with radiogenic isotopes, and is of so-called J-type, in contrast with Moorbath's interpretation on the basis of normal leads. Mitchell and Krouse therefore deny that the lead ratios can be used as age indices, maintaining that they merely reflect differences in primary sources from field to field. The whole subject deserves, and is receiving, further attention.

Meanwhile uncontroversial evidence of the age of deposition of uranium minerals in Cornwall is forthcoming from the application of the uranium–lead method by A. G. Darnley and others (1965). It was concluded that there were at least three periods of primary uranium mineralization, at about 290 m.y. (Permo-Carboniferous), 225 m.y. (Triassic) and about 50 m.y. (Lower Tertiary); there may be others, for example Lower Cretaceous. The possibility of some regeneration of the deposits was considered, but S. H. U. Bowie (in discussion on Dunham *et al.* 1965) commenting on the situation has strongly maintained that the evidence must be accepted as showing that in Cornubia the filling of the veins was either continued over very long periods or else was a recurrent process.

Up to that time there had been an understandable tendency among metalliferous geologists to think of the formation of an ore deposit as a short-period event, comparable in its time-scale with the intrusion of a dyke or sill. Some, like Finlayson, had found themselves obliged to postulate continuing mineral deposition, for example in the Mendips, from the Hercynian orogeny in the Carboniferous–Permian interval at least to Middle Jurassic times. As already noted, there is very clear evidence that lead, zinc, fluorine, and barium were still available in quantity in north-east England well into the Permian.

More recently another approach to the problem has been tried out. As Finlayson (1910*b*) first demonstrated in the cases of Rotherhope Fell (Whin dolerite), Leadhills (slate), Conway (ash), and Foxdale (granite), the wall-rock alteration accompanying the formation of metalliferous veins involves appreciable addition of potash, generally expressed as sericitization (but now known to involve the formation of other clay-micas). The new micas can be separated and subjected to ^{40}Ar–^{39}Ar and K–Ar procedures. This was tried out on material collected by P. R. Ineson from altered dolerite at Closehouse and Settlingshouse mines in the northern Pennines, the determination being made by J. A. Miller and J. G. Mitchell on the omegatron at Cambridge (Dunham *et al.* 1968). The results indicated one initial mineralization of Hercynian date, 284 ± 40 m.y., but also pointed to recurrences at 230 m.y. (Permian) and 170 m.y. (Jurassic) dates. These results have been found un-acceptable by R. H. Mitchell and Krouse (1971) on the grounds that the fine grain-size of the micas and the separation method would be likely to lead to loss of argon. Subsequently, Ineson and J. G. Mitchell (1974) have analysed argillaceous alteration zones, related to the Ochil fault and the veins at Tyndrum, Wanlockhead, and Strontian, obtaining a spread of apparent K–Ar ages from Viséan (Wanlockhead) to Permian (Strontian). In the case of Strontian (Gallagher et al. 1971), model lead ages by Darnley showed, in addition to an event in Permian times, a second one in Tertiary times.

Of all the available data, there is no doubt that the most satisfactory evidence comes from U–Th–Pb determinations. These may be regarded as indicating that the emplacement of metalliferous deposits may not be a short-lived effect, but one that may span several epochs, or recur repeatedly. The fact is that in epigenetic ore deposits we are seeing the results of the filling, reopening, and renewed filling of channels for the movements of fluids in the earth's crust which, once formed, were often never again sufficiently sealed up by deposits to render them completely impermeable. As will appear later, it is especially the case with deposits formed at temperatures below 200 °C that the evidence of age is conflicting, and the affiliation to a major orogenic or plate-margin event is not necessarily strong.

4. Structural control of oreshoots

The dating of British mineral fields, outside Cornwall and Devon, thus presents many unsolved problems, but it is no mere matter of academic interest, for it affects thinking about where to seek new fields and new deposits. More amenable to the industrial geologist is the question of the physical factors that have localized the ores. Regional stress patterns are of primary importance, for the fissures that have become veins are to be regarded as the strain-effects of such patterns. The close folding—dominantly isoclinical in Wales, the Lake District and the Southern Uplands, recumbent in the southern part of the Highlands—which formed the Caledonides at the maximum of the orogeny was not well calculated to produce tensile fractures. Tensile fracture certainly did follow the post-tectonic 'newer' granites but many of these fractures were quickly sealed up with dykes and sills. Tensile fracturing in the Plynlimon field was held by O. T. Jones (1922) to have been contemporaneous with, or to have followed soon after the folding. Very likely the position of the small deposits round the Harlech Dome, which exposes the type-area of Cambrian in Wales, were in some way related to the margins of that structure, as Wheatley (1971) has suggested. In Cornwall and Devon, tensile fracturing along ENE–WSW to E–W directions set in as the batholith consolidated, and the first set of fractures received granite-porphyry dykes ('elvans' in Cornish parlance) probably intruded when the deep interior of the batholith still contained fluid magma. The ENE–WSW to E–W direction continued to be the major strain direction and fractures later in

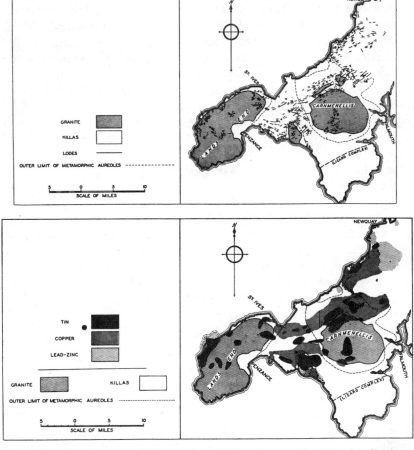

Fig. 7.3. Sketch-maps of Cornwall and West Devon, showing the distribution of mineral lodes, principal non-ferrous metals and granites by H. G. Dines. (Reproduced, by permission, from *The metalliferous mining region of south-west England, Mem. Geol. Surv.*, 1956, Figs. 4a, 4b.)

date than the dykes became the mineral veins (Fig. 7.3). These lie along the main axis of the batholith (see Hosking 1964). They dip at an average of about 75° both north and south, but there are a few like the Great Flat Lode at much lower inclinations. At the new Wheal Jane Mine, the main lode is a strong shear zone developed beneath the forewall of a dyke dipping at 35° W (Rayment *et al.* 1971). The initial set are transversed by a later set, more or less at right-angles, which heaves them; however, near Land's End the main productive veins run NNW–SSE. In the northern Pennines the Alston Block is cut by over 500 veins, arranged in a highly geometrical pattern of ENE-, NNW-, and WNW-trending fractures which I believe to have been the response of the brittle rocks of the Yoredale facies to the gentle Permo-Carboniferous pressures that

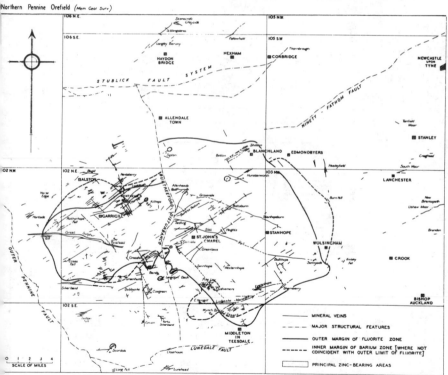

Northern Pennine Orefield *(Mem Geol Surv.)*

Fig. 7.4. Sketch-map of the northern part of the Northern Pennine orefield, to show the distribution of mineral veins, and the fluorite and barium zones by K. C. Dunham. (Reproduced, by permission, from *Geology of the Northern Pennine orefield, Mem. Geol. Surv.*, 1949, Plate II.)

151

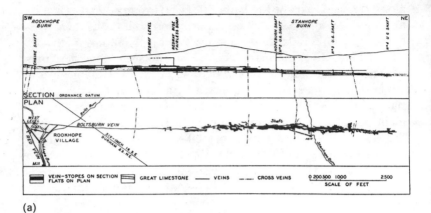

(a)

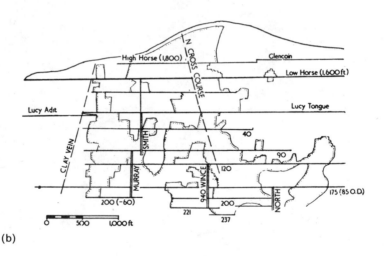

(b)

Fig. 7.5. Comparison of longitudinal sections of oreshoots. (a) Boltsburn Mine, Northern Pennines, by K. C. Dunham; (b) Greenside, Lake District, by T. Eastwood; (c) Snailbeach, west Shropshire, by H. Dewey and B. Smith; (d) Idealized cross-section of North Pennine fissure vein. (Reproduced by permission, (a) from *Geology of the Northern Pennine Orefield, Mem. Geol. Surv.*, Fig. 21; (b) from *Future of non-ferrous mining in Britain and Ireland*, 1958, Fig. 1, p. 152; (c) from *Lead and zinc ores in pre-Carboniferous rocks of west Shropshire and North Wales, Spec. Rpt. Min. Res., Geol. Surv. G. B.*, 1922, **23**, Fig. 10; (d) from *Geology of the Northern Pennine Orefield*, Fig. 9.)

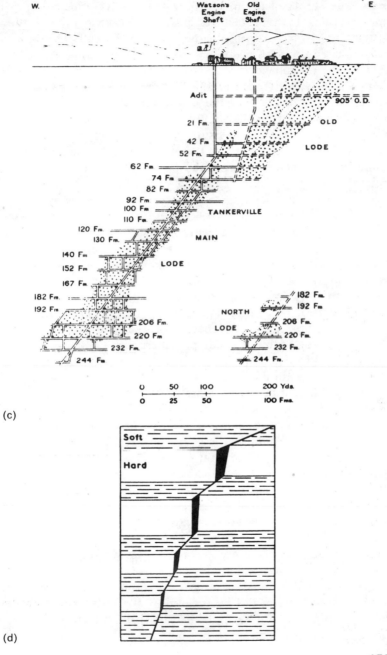

(c)

(d)

153

raised the Teesdale half-dome (Fig. 7.4). In all the principal British mineral fields, the vein-fractures must be seen first in their setting within the regional structural pattern. The ENE trend, for example, was initiated during Caledonian movements in northern England and southern Scotland (notice, for instance, how the Geological Survey 1 inch to 10-mile geological map shows a great swarm of dykes in this direction in the south-west Scotland part of the Caledonides); but it is reflected precisely in the Lower Carboniferous rocks of the Alston Block which rest on a basement of Lower Palaeozoic rocks. On the other hand, at Leadhills, in the Lake District, and at Llanwrst, the generally N–S trend of the vein fractures cuts across the Caledonoid grain (Table 7.1).

Notwithstanding the lithostatic pressure at the depths (probably 1–5 km) below surface at which the British non-ferrous deposits formed, tensile or transcurrent movements along the fractures were evidently sufficient to provide new open space to permit ingress of fluids. The further development and ultimate filling with the products of crystallization of the mineralizing fluids produced bodies of ore (technically known as oreshoots) that are nearly always crudely lenticular in form. The simplest case is where the sides of the fracture are non-planar, and the effect of the fault movement was to bring small protuberances on the walls into contact so as to give support to an open channel. A more advanced case is where, as in the Pennines, the fractures cut across beds of markedly different physical properties, so that they are 'refracted' from one bed to the next (Fig. 7.5). Coupled with normal fault movements of small amounts (generally less than 10 m), open spaces are created in the brittle beds where the fractures are nearly vertical. In the third dimension, the open spaces coincided with hard, brittle beds on the vein-walls, and since these beds are thin (the thickest individual one is little more than 12 m thick), the effect was to create a network of shallow, laterally extensive permeable channels. The corresponding oreshoots (Fig. 7.5a) have been described as ribbon-shaped (Dunham 1949).

Contrasted with these are the oreshoots in Lower Palaeozoic strata, where the physical characteristics of the wallrocks remain constant for substantial depths; the oreshoots that result are roughly equidimensional. The Greenside (Lake District) oreshoot depicted in Fig. 7.5b, for example, had hard Borrowdale pyroclastic rocks as wallrock from top to bottom and perhaps lay in the limb of a steep fold; it died out in depth where the fracture passed into soft Skiddaw Slates. The plunging oreshoot from Shropshire illustrated in Fig. 7.5c is related to a steeply dipping favourable bed, the Mytton Grit. The wallrocks, for purely physical reasons, affect the distribution of oreshoots; when favourable, they are known as 'bearing beds' (Table 7.1) but this by no means implies that the local wallrock was also the *source* of the fluids. We shall return to this point later. Where the wallrocks are fairly homogeneous there are still likely to be variations in width of oreshoot resulting from minor curvature or variation in the extent of opening of the fracture. Good examples can be seen in the accounts by Garnett (1966) and Taylor (1966) of the oreshoots at the Geevor and South Crofty tin mines respectively, where they used the device of contouring in relation to an assured plane (known as the Connelly plane after the Australian geologist who devised it).

Even where they are related to bearing beds, oreshoots do not continue indefinitely along the strike; the fracture closes, or dies out. Often though not always, this occurs near or at the intersection of another fracture. Such a 'cross-course' may be post-mineralization in age, in which case the shifted continuation of the oreshoot beyond it must be sought, but in many cases the fractures are contemporaneous and there is no shifted continuation to be found.

As a result of recent work it has become likely that the development of open space along mineral veins is not solely the result of different rock pressures developed in connection with tensile faulting but is also related to fluid pressures. Experiments with injection of waste fluids into deep boreholes have shown that modest pressures, often little more than that exerted by the mass of the column of fluid in the borehole, are sufficient to break the rocks at depths of 3–4 km, and that fracturing, once started, tends to be propogated spasmodically. The recent case at Denver, Colorado is well known; the injection process set off a long series of minor earthquake shocks (see p. 301). W. J. Phillips (1972) in an important contribution has analysed the effect of introducing hydrothermal fluids into a developing fracture system and concludes that not only would the propagation of the fractures be stimulated, but wide zones of breccia would be produced. Such zones are well known, not only in the Plynlimon field, from which Phillips draws his illustrations, but in most British orefields. He suggests that tensile

155

fracturing releases pore-waters from sediments, contributing to the formation of hydrothermal fluids. When the sheets of these exceed a critical size, hydraulic pressures transmitted from below cause brecciation and propagation of the upward edge of the fracture zone, in turn profoundly affecting the migration paths of the fluids.

Table 7.1 brings out the shallow depth within the crust to which the British orefields have so far been explored. Only in Cornwall has a depth of 1 km been exceeded; few mines elsewhere are more than 0·5 km deep. In some mines, perhaps in most, the oreshoots have died out in depth and it has been too costly to look for others; in others the workings became uneconomic with increasing depth and greater volumes of water to pump. In many fields it is nevertheless true to say that there is no geological reason, unless it is the critical depth at which hydraulic fracturing becomes possible, why oreshoots should not exist at depths considerably greater than 1 km.

5. Temperature control of oreshoots

Evidence that mineralization in Cornwall was brought about by hot, saline waters was first obtained 115 years ago by that remarkable metallurgist and petrographer, Henry Clifton Sorby (1858), who observed that drops of liquid were enclosed inside quartz crystals from the veins. The liquid, considered by him to have been included as the crystals grew, was insufficient to fill the cavities, but he suggested that on heating it would expand to fill the space at the temperature at which the quartz crystallized. His measurements, carried out on thick sections under the microscope, enabled him to calculate temperatures of about 200 °C at assumed water-vapour pressures. His results were received at the Geological Society with scepticism, and it was not until 1930 that they were followed up by experimental determination of temperature at which the liquid exactly filled the cavity. Many determinations established that the method is a valid one, and that we may accept that the Cornish veins were filled by hot solutions which cooled as they travelled upwards and outwards from their sources.

It has long been established that consistent variations in the contents of many veins occur as they are followed downwards. Dolcoath, the deepest of Cornish mines, furnishes a fine example (Fig. 7.6). From the surface approximately down to the 200-fathom level (372 m) the vein was worked for copper; beneath an intermediate belt, the filling changed in character so that from the

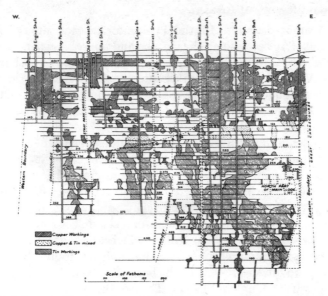

Fig. 7.6. Longitudinal section of Dolcoath Mine, West Cornwall, to show mineral zoning by H. G. Dines. (Reproduced, by permission, from *The metalliferous mining region of south-west England, Mem. Geol. Surv.* 1956, Fig. 17.)

200-fathom level to the bottom at 1006 m, the ore yielded cassiterite (SnO_2), instead of chalcopyrite ($CuFeS_2$). The explanation is that the vein was acting as a natural fractionating column, so that minerals which become insoluble at high temperature formed in the lower part of the vein, while those in the upper part crystallized at lower temperatures. The phenomenon is known as mineral zoning, and on the basis of observations in many different mines a succession of zones has been built up. The summary by H. Dewey (1925) gives a simple picture, but much fuller details are to be found in Hosking (1964) and in Edmonds *et al.* (1969). There is a marked tendency for the upper, cooler zones to spread laterally, so that each higher one becomes wider (Dines 1956). Some points on the temperature-scale have been established by modern fluid inclusion studies carried out by F. J. Sawkins (1966), who showed, on quartz contemporaneous with, or a little later in crystallization than, cassiterite, a range from 440 °C down to 298 °C; quartz and fluorite associated with chalcopyrite 256–48 °C and 256–70 °C respectively; on sphalerite (ZnS), 150–70 °C; and on fluorites post-dating galena, 115–46 °C.

Mineral zoning essentially in a vertical sense is known from

157

several other orefields. In the Plynlimon district, many of the oreshoots were becoming richer in zinc and poorer in galena as they were followed downwards; there were similar cases at Llanwrst, and in West Shropshire. In the latter district, several lead veins contained abundant barite ($BaSO_4$) in the higher levels, but this diminished or died out downwards. The same was true at Greenside and Force Crag in the Lake District.

The laterally extensive channels of fluid movement in the northern Pennines present a different case. Here, while there are instances of zoning in a vertical sense, the lateral change of mineral contents in the vein is far more obvious. The highest-temperature zones carry fluorite, quartz, and chalcopyrite but are of slight economic importance and are interesting chiefly because they form foci in broad fluorite–galena zones; which are concentrically surrounded by a region where barium minerals are found, but where fluorite is no longer present. In the barium region, the sulphides die out outwards (Dunham 1949). The southern Pennines show a comparable zonal pattern, but here the fluorite zone parallels the eastern Edale Shale/Limestone contact, giving place westward to barite or calcite; the latest review of the subject is by T. D. Ford and P. R. Ineson (1971). Temperatures can be assigned from the investigations by F. J. Sawkins (1966) and E. Roedder (1969) for the northern and southern fields respectively. The fluorite zone in the Alston Block area ranges from 220 °C to 100 °C, with a reasonably continuous outward change; barium minerals appear to have formed at temperatures down to normal. Fluorite temperatures in Derbyshire are in the range of 70–140 °C.

It may reasonably be claimed that the lateral zonal system in the Pennines corresponds with the outer zones of the Cornwall–Devon area, particularly as regards the succession Cu–Zn–Pb and fluorite–barite (Fig. 7.7). The hypothesis that granite of post-Carboniferous age lay concealed beneath the Carboniferous rocks of the northern Pennines therefore seemed worth testing, especially after the demonstration by Bott and Masson-Smith (1957) that the gravity field of that area could be interpreted satisfactorily only if granitic stocks underlay the two main centres of the fluorite zone. In 1960–1 boring was undertaken at Rookhope, County Durham to investigate what was already being called the 'Weardale granite'. The geophysical results were fully substantiated, but the granite proved to be of Caledonian, not Armorican, age (Dunham *et al.*

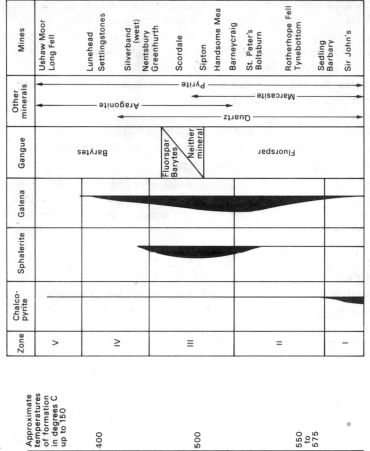

Fig. 7.7. Comparison of mineral zoning in Cornwall–Devon by H. Dewey; (b) the Northern Pennines by K. C. Dunham. (Reproduced, by permission, (a) from *Proc. Geol. Assoc.* 1925, **36**, Fig. 7, p. 128; (b) from *Q. Jl. geol. Soc. Lond.* 1934, **90**, Fig. 2, p. 705.)

1965), with a weathered zone separating it from unmetamorphosed basal Carboniferous strata. Nevertheless, this granite clearly played some part in localizing the mineral field, just as it seems to have some connections with the devolatilization of coal seams in a westward direction from the Durham coalfields. Fractures which could have been part of the feeder system for the Pennine ores continue down into it, and it may thus have acted as a structural channel. Its low density probably accounts for the doming of the Carboniferous sediments; but most significant of all, there is still an abnormally high heat flow through it, according to Bott and others (1973), twice as high as could be explained by the radioactive minerals it contains. It is interesting also to note that the Cornubian batholith has a high heat flow as indicated by the rapid downward increase of temperature in mines, and the many hot springs found in them during the past century. Surprisingly, therefore, it now appears that two of the major British mineral fields, where it may reasonably be claimed that the main surge of metallization took place at least 250 m.y. ago, are still areas of relatively high heat flow. Hot waters were found in oil borings east of the north Derbyshire mineral field but at one key point (Eyam) beneath it there appears to be evidence of a flow of cold water. For the remaining mineral districts we have no satisfactory evidence of the present thermal regime.

6. Mineralizing fluids and metal sources

It has already been mentioned that H. C. Sorby's observation on Cornish vein quartz showed that it enclosed water with potassium and sodium chlorides in solution, some sulphates were present and there was an acid reaction. This observation, also long neglected, has been pursued during the past few decades; it is particularly important because the fluid must be regarded (provided that no later ingress has occurred) as a sample of the actual mineralizing solution. Only a few such fluids have yet been fully analysed, but there are many partial analyses giving total salinity (based on freezing temperatures, observed under the microscope) and determinations of Na/K ratios. For Cornwall, Sawkins (1966) found a wide range of salinities, from 2·0 to about 50 equivalent weight per cent NaCl, and Na/K ratios varying from 2·8 to 17·9. In a more detailed study by the same author of northern Pennine minerals, the salinities of the enclosed solutions were found to range from 20 to 23 per cent NaCl-equivalent, or about six times the salinity of ocean water. Though far more evidence is needed, there is a strong indication that the mineralizing fluids were, in fact, hypersaline brines. Chemically this is a satisfactory hypothesis, since chloride complexes of the nonferrous metals are sufficiently soluble to enable them to be transported.

Such saline solutions could have originated in several different ways. The obvious comparison is with the deep formation waters encountered by any borehole drilled beyond about 600 m depth (Dunham 1970). Oil-well drilling has made these familiar as the edge-waters of oilfields. They may arise from the burial of sea-water with the enclosing sediments (connate water) followed by filtration through semipermeable membranes like clay or shale beds, which allow the water to pass, but not the larger ions. Alternatively they may result from solution of salt deposits in the stratigraphic section, but this is less easy at depth than might at first appear, owing to the strong tendency in salt deposits to seal up any fractures. For such waters to attain temperatures corresponding with the central zone in the northern Pennines (220 °C) they would have to have descended to at least 7 km depth before returning towards the surface, assuming the present temperature gradient; but this is by no means improbable, since oil wells at depths greater than 7 km continue to encounter hypersaline formation waters.

Nor should it be supposed that highly permeable reservoirs are necessarily involved. The small relative amounts of pore-water eliminated from shales and greywackes could, as Phillips (1972) has stressed, also make an effective contribution to the generation of hydrothermal solutions.

If the formation waters migrating upward from large bodies of sediment are indeed the mineralizers, the sources of the metals and other elements concentrated in the ore deposits must be in the rocks, sedimentary, metamorphic, or igneous, through which the fluids pass on their way to the vein-fractures. Figures for the Yorkshire Pennines (Dunham 1959) suggest that dissolution of metals from a shallow layer within the vein-walls is not adequate; it would be necessary to have large volumes of rock leached at some stage.

The alternative is to suppose that the fluids are the residua from crystallizing igneous magmas ('wet' magmas, that have contained appreciable amounts of water in solution). The evidence of hot springs and formations found in volcanic regions suggests that these, too, may be hypersaline brines. In this case, the magma could be the source of the metals; but it should be noted that as magmas crystallize, lead and sometimes barium tend to go into feldspars, zinc and barium into dark micas, zinc into amphiboles; not all the metal content of a magma can be expected to be available for residual concentration in the last watery brines. Such brines are also powerful solvents; however, they will only be effective if they pass through large volumes of rock. Perhaps a sharp distinction ought not to be drawn between formation and igneous-residual waters; for example, it has been suggested that a crystallizing granite may suck into itself formation waters from the sediments it has intruded, only to expel these as hydrothermal solutions when crystallization is coming to an end. It is also certain that, in some cases, hot magmas do no more than provide the heat to drive a convective system involving water of surface origin. White (1969) remarks that at least 19 huge geyser-bearing systems have now been explored by drilling, demonstrating that meteoric water can penetrate at least to 3 km depth, even in crystalline rocks of low permeability.

It may finally be noted that study of stable isotopes, for example of sulphur and of oxygen, may give clues to the mineralizing fluids and to metal sources. It has generally been accepted that where the ratio $^{22}S/^{24}S$ corresponds closely with that of the sulphur of meteorites, the sulphur is primeval and thus igneous in parentage; an apparent example is to be found in the work by Mitchell and Krouse (1971) on Greenhow, Yorkshire. However, when a wider field of the Pennines is surveyed (Solomon *et al.* 1971), the range in sulphide sulphur isotopes is so great as not to be characteristic of magmatic deposits or of derivation from the enclosing Carboniferous sediments. Only far-travelled formation waters, mixing in the barium zone with local groundwaters, appear to offer an adequate explanation. The whole subject is, however, as yet, in its infancy, offering promise for the future.

The important questions of the concentration of metals during sedimentation or evaporation of the sea have not been enlarged on here since, although it is admitted that such processes may contribute to pre-concentration, it is not thought that they were of major importance in generating any of the British ore deposits other than the Mesozoic sedimentary ironstones. Thus, although the

Carboniferous Limestone districts contain nonferrous and ferrous ores that are broadly stratiform, in all instances these can be shown to be related to feeding fissures, along which the mineralizing fluids were introduced. The stratiform deposits in these cases are produced by replacement of the limestone.

In conclusion, therefore, it is suggested that the metalliferous deposits of Britain (other than the Mesozoic ironstones) are, with hardly any exceptions, epigenetic (later than the enclosing rocks) and closely linked with Palaeozoic events, mainly but not entirely with the later part of that era. But direct igneous affiliation seems strongly established only in the Armorican subprovince and at Carrock Fell. Elsewhere the range in apparent ages and absence of demonstrable igneous affiliations favours deep groundwater as the chief collector and provider of the deposits, being at its most effective where abnormal heat flows obtained or now exist. Were these sites of very deep igneous activity? Perhaps when we investigate seriously the lower crust (rather than the upper mantle) the answer will emerge.

References

BOTT, M. H. P., and MASSON SMITH, D. (1957). The geological interpretation of a gravity survey of the Alston Block. *Q. Jl. geol. Soc. Lond.* **113**, 93–117.

—— JOHNSON, G. A. L., MANSFIELD, J., and WHEILDON, J. W. (1973). Terrestrial heat flow in north-east England. *Geophys. J. R. Astr. Soc.* **27**, 277–88.

DARNLEY, A. G., ENGLISH, T. H., SPRAKE, O., PREECE, E. R., and AVERY, D. (1965). Ages of uraninite and coffinite from south-west England. *Min. Mag.* **34**, 159–76.

DE LAUNAY, L. (1905). *Formation de gites métallifères ou métallogenie.* (2nd edn.) Paris.

—— (1913). *Métallogenie.* Paris.

DEWEY, H. (1925). The mineral zones of Cornwall. *Proc. Geol. Ass.* **36**, 107–35.

DINES, H. G. (1956). The metalliferous mining region of south-west England. *Mem. geol. Surv. U.K.* 2 vols.

DUNHAM, K. C. (1949). Geology of the Northern Pennine orefield. Vol. I. Tyne to Stainmore. *Mem. geol. Surv. U.K.*

—— (1952*a*). Fluorspar. *Mem. geol. Surv. Spec. Rp. Miner. Resour. Gt. Br.* 4th edn.

—— (1952*b*). Age-relations of the epigenetic mineral deposits of Britain. *Trans. geol. Soc. Glasg.* **21**, 395–429.

—— (1959). Epigenetic mineralization in Yorkshire. *Proc. Yorks. geol. Soc.* **32**, 1–29.

DUNHAM, K. C. (1970). Mineralization by deep formation waters. A review. *Trans. Instn Min. Metall.* **79**, B127–36.

—— DUNHAM, A. C., HODGE, B. L., and JOHNSON, G. A. L. (1965). Granite beneath Viséan sediments with mineralization at Rookhope, northern Pennines. *Q. Jl. geol. Soc. Lond.* **121**, 383–417.

—— FITCH, F. J., INESON, P. R., MILLER, J. A., and MITCHELL, J. G. (1968). The geochronological significance of argon–40/argon–39 age determinations on White whin from the northern Pennine orefield. *Proc. R. Soc. A.* **307**, 251–66.

—— and POORE, M. E. D. (organizers) (1974). A discussion on the exploitation of British mineral resources (other than coal and hydrocarbons) in relation to countryside conservation. *Proc. R. Soc. A.* 339.

EDMONDS, E. A., McKEOWN, M. C., and WILLIAMS, M. (1969). South-west England. *Br. reg. Geol.* HMSO.

FINLAYSON, A. M. (1910*a*). The metallogeny of the British Isles. *Q. Jl. geol. Soc. Lond.* **66**, 281–98.

—— (1910*b*). Problems of ore-deposition in the lead and zinc veins of Great Britain. *Q. Jl. geol. Soc. Lond.* **66**, 299–378.

FORD, T. D., and INESON, P. R. (1971). The fluorspar mining potential of the Derbyshire orefield. *Trans. Instn Min. Metall.* **80**, B186–210.

GALLAGHER, M. J., MICHIE, V. McL., SMITH, R. T., and HAYNES, L. (1971). New evidence of uranium and other mineralization in Scotland. *Trans. Instn Min. Metall.* **80**, B150–73.

GARNETT, R. H. T. (1966). Relationship between tin content and structure of lodes at Geevor Mine, Cornwall. *Trans. Instn Min. Metall.* **75**, B35–50.

HOSKING, K. F. G., and SHRIMPTON, G. J. (eds.) (1964). *Present views of some aspects of the geology of Cornwall and Devon.* Penzance.

INESON, P. R., and MITCHELL, J. G. (1974). K–Ar isotope age determinations from some Scottish mineral localities. *Trans. Instn Min. Metall.* **83**, B13–18.

INSTITUTION OF MINING AND METALLURGY (1958). *The future of non-ferrous mining in Great Britain and Ireland.* London.

JONES, O. T. (1922). Lead and zinc. The mining district of north Cardiganshire and west Montgomeryshire. *Mem. geol. Surv. spec. Rep. Miner. Resour. Gt Br.* **20**.

MOORBATH, S. (1962). Lead isotope abundance studies on mineral occurrences in the British Isles and their geological significance. *Phil. Trans. R. Soc. A.* **254**, 296–360.

MITCHELL, R. H., and KROUSE, H. R. (1971). Isotope composition of sulphur and lead in galena from the Greenhow–Skyreholme area, Yorkshire, England. *Econ. Geol.* **66**, 243–51.

MORGAN, D. A. O. (1974). Metalliferous potential of the United Kingdom. *Proc. R. Soc. A.* **339**, 289–98.

PHILLIPS, W. J. (1972). Hydraulic fracturing and mineralization. *Q. Jl. geol. Soc. Lond.* **128**, 337–60.

RAYMENT, B. D., DAVIS, G. R., and WILLSON, J. D. (1971). Controls to mineralization, at Wheal Jane, Cornwall. *Trans. Instn Min. Metall.* **80**, B224–37.

ROEDDER, E. (1969) In *Sedimentary ores, ancient and modern* (ed. T. D. Ford), pp. 87, 92, Leicester.

SAWKINS, F. J. (1966). Preliminary fluid inclusion studies of the mineralization associated with the Hercynian granites of south-west England. *Trans. Instn Min. Metall.* **75**, B109–12, B242, B307.

SOLOMON, M., RAFTER, T. A., and DUNHAM, K. C. (1971). Sulphur and oxygen isotope studies in the northern Pennines in relation to ore genesis. *Trans. Instn Min. Metall.* **80**, B259–75; **81**, B172–8.

SORBY, H. C. (1858). The microscopical structure of crystals, indicating the origin of minerals and rocks. *Q. Jl. geol. Soc. Lond.* **14**, 453–500.

SUESS, E. (1909). *Das Antiltz der Erde*. Vienna.

TAYLOR, R. G. (1966). Distribution and deposition of cassiterite at south Crofty mine, Cornwall. *Trans. Instn Min. Metall.* **75**, B35–49, B172–6.

WHEATLEY, C. J. V. (1971). Aspects of metallogenesis within the Southern Caledonides of Great Britain and Ireland. *Trans. Instn Min. Metall.* **80**, B211–23.

WHITE, D. E. (1969). Thermal mineral waters of the United States—brief review of possible origins. *Rep. 23rd Int. geol. Congr.* **19**, 262–86.

8. Aggregates, sand, gravel, and constructional stone

1. Introduction

The abstraction of bulk constructional materials is one of the principal industries of the United Kingdom and the U.K. tonnage alone almost equals the world production of steel. Over the period 1970–3 an annual average of 117.5×10^6 tonnes of sand and gravel, principally for concrete, were extracted in England and Wales, 106.8×10^6 tonnes of crushed rock aggregate, quite apart from building stone, were quarried, and during 1971 alone 60×10^6 tonnes of shale and limestone were used in cement manufacture. The increasing demand for these materials is caused both by population growth and industrial expansion and a constant demand for new housing, offices, motorways, airports, public utilities (e.g. hospitals, schools, dams, harbours), and all the other man-made construction which characterizes a dominantly urban industrial society.

Fortunately most natural bulk materials are cheap: they are plentiful, relatively easy to extract (hardly any are mined), and require a minimum of processing before sale. Sands and gravels, for example, usually require only washing to remove clay coatings and screening, and have a general on-site value of £1 per tonne (1975). Crushed rock aggregate (e.g. rip-rap, road aggregate), which is slightly more expensive to extract and process, has an on-site value of approximately £1.50 per tonne. Massive rock if trimmed or polished costs £20–25 per square metre (Open University 1974). Table 8.1 outlines the main derivations of bulk materials, their utilization, processing, and on-site value.

The potential resources of bulk aggregate in Great Britain are theoretically almost unlimited. The estimated resources are 290×10^9 tonnes of sand and gravel, and 7300×10^9 tonnes of hard rock respectively (Verney 1976). In practice, both quarrying and

TABLE 8.1

Sources and utilization of natural constructional aggregate

Raw material	Main type	Utilization	Method of extraction	Processing	Approximate waste (%)	On-site value
Rock	Granite Sandstone Basalt Marble Slate	Building stone Road aggregate Rip-rap Rock fill	Quarrying	Trimming for building stone Crushing for rip-rap and aggregate	50–59 Nil	£20–25/m² £1.50/tonne
	Limestone	Cement	Some softer rocks—(mechanical excavation)	Grinding of limestone for cement	Nil	£9.50/tonne
Sand and gravel	Fluvial, estuarine, glacial, and marine deposits	Mainly concrete aggregate	Mechanical excavation	Grinding	< 5	£1.00/tonne
	Unconsolidated and poorly cemented sediments	Moulding sands Glass sands Refractory bricks				
Clay	Clay or shale	Cement	Mechanical excavation	Grinding, moulding		
	Brick clay	Bricks and tiles		Kilning	< 3	£13.28/100 bricks
	China clay	Pottery, paper, paint thickener	Hydraulic washing	Washing	60	£14.00/tonne

sand and gravel extraction are controlled in order to avoid the loss of agricultural or building land or the despoilation of an amenity area. The demand placed on the extractive industry is enormous but at the same time the demand fluctuates markedly with changes in local government policy and in politics at a national level.

The principal sources and utilization of constructional materials (including industrial wastes) will be discussed in this chapter.

2. Sand and gravel

2.1. General

The bulk of sand and gravel which is extracted is supplied to the concrete industry for building. In 1971, for example, 110 million tonnes of sand and gravel were extracted in England and Wales (Open University 1974). Figures for the United States show equivalent large tonnages; in 1956, 620 billion tonnes of sand and gravel were sold there (Lenhart 1962). Continual expansion in house-building and in the construction of offices, motorways, industrial sites, etc., demands an ever-increasing expansion in the production of sand and gravel, and a corresponding survey of natural resources of these materials has been implemented. The summary below gives an indication of the average requirements for certain specifications.

	Aggregate (tonnes)
Three-bedroom detached house	50
15-floor office block	2 250
1 km of motorway	62 500
1 km of class A road	12 500
Airport runway (5·5 km × 50 m)	270 000

The production of sand and gravel has now exceeded the extraction of coal so that the extraction of sand and gravel has become the principal extractive industry in the British Isles. It is therefore responsible for considerable exploitation of our natural resources. Each year 1700 ha of land in England and Wales is opened for sand and gravel extraction with an ensuing loss both of open space and land available for building. In south-east England, and especially in the Greater London area, where most sand and gravel pits are situated, the demand for aggregate and the demand for building land conflict with each other and yet so long as population and industrial expansion continue in the area there seems no solution to this problem other than by bulk importation to the region. The sand and gravel industry does, however, have a good record for restoring worked-out pits. In the Thames Valley, most gravel pits are excavated to below the water table and the ensuing lakes can relatively easily be converted to nature reserves for wildfowl, developed for angling, converted to yacht marinas, or filled in and landscaped to form attractive open 'leisure' areas (SAGA 1967).

Sand and gravel are clastic sedimentary rocks requiring a medium for their transport and thus reflecting their environment of deposition. These materials are found in alluvial fans, river terrace, deposits, aggregations of flood-plain material, sand dunes, screes, talus, alluvial deltas, loess, moraines, tills, and glacial outwash materials. Sand and gravels also arise from the weathering *in situ* of igneous and older sedimentary rocks, but such deposits are commoner in tropical areas where ground-water leaching and alteration are important weathering processes. Tuffs and agglomerates form additional sources of sand and gravel although the clay mineral content and the presence of chalcedonic quartz and zeolites generally renders them unsuitable for concrete. Poorly cemented sandstones, such as occur in the Bunter and Cretaceous, are also worked for sand and gravel.

The environment of deposition of gravels is reflected by the degree of sorting, rounding, pebble size, etc. Well-sorted gravels, free of clayey or silty 'fines', are preferred to the poorly sorted gravels and sands such as are associated with moraines and glacial outwash deposits, which often contain abundant fines and are either unsuitable for concrete aggregate or require expensive benefication treatment.

In the geological assessment of any deposit of sand and gravel in the field, the following points are especially important (Krynine and Judd 1957).

Size of deposit. Field examination will indicate the volume, lateral extent, structural conditions, depth of overburden to be removed, and other characteristics of the deposit. Small deposits which will rapidly be exhausted are normally avoided unless they are required for a specific local project. The duplication of plant over several small sites is normally avoided, for it is expensive and increases the total cost.

Aggregates, sand, gravel, and constructional stone

Location of deposit. The deposit should be sited as near the market or particular project as possible and the site should be easy of access. Transport over long distances increases the aggregate costs. In south-east England, for example, gravel pits in the Thames Valley which work flood-plain and Pleistocene terrace gravels are near to an insatiable market and have easy and rapid access to it; materials costs are thus theoretically at a minimum. In some instances it may be cheaper to beneficate a less suitable local aggregate rather than transport a good aggregate over a longer distance (Lenhart 1962).

Groundwater conditions. The level of the water table is often important since aggregate dredged from below water is more expensive to raise than aggregate which is mechanically dug in the dry. Sudden flooding of the pit by a rise in the water table may also damage expensive plant. Nevertheless, the bulk of British gravels are dredged either from below the water table (as in the Thames Valley) or from estuaries and offshore deposits.

In the treatment of sand and gravel, abundant water must be available for washing to remove clay fines, clay coatings, salts, etc. In the British Isles this need not present specific problems. Sea-water can be used to wash gravel, although salts may subsequently have a deleterious affect on concrete made from sea-washed aggregate.

Deleterious constituents. An on-site examination of the rock fragments making up the aggregate can determine its suitability for use either as concrete or road aggregate. Iron pyrites, coal, mica, shale, micaceous rocks (e.g. schists), and laminated flaky or elongated particles and organic impurities should be absent (BS 1199). Clay, silt, and fine sand should not exceed 5 per cent. Other deleterious constituents (chert, opaline silica, etc.), may be identified during a petrological examination. In the laboratory the aggregate can be subject to a variety of physical tests according to its final use; e.g. sands intended for plastering will have specifications somewhat different from those required for concrete or moulding sands. (For specific tests see BSS 812, 410, 1198, and 1200.)

For concrete aggregate the geologist will normally be expected to evaluate the following criteria:

(i) The proportion of coarse and fine material present. This is normally estimated using a standard range of British sieves (B.S. 812 and B.S. 410) and from as large and as typical a sample as possible.

(ii) Grading: Although a percentage of fines in excess of 5 per cent is to be avoided (B.S. 1199), a predominance of one size of aggregate particle may adversely affect the workability of concrete. Numerous voids may then occur between the particles and in order to fill them the proportion of cement in the mix must be increased; the resulting concrete may be expensive, low in strength, or both.

(iii) Surface texture. Smooth pebbles are not conducive to good cement for the bond between aggregate and cement is weak. Similarly, aggregate with numerous small pores also form poor structural concrete since the pores permit the capillary movement of water and in cold climates the freezing of this water causes expansion and consequent failure of the concrete. Aggregate with few pores does however produce good aggregate for hard-wearing pavements since they are resistant to abrasion. Quartz, quartzite, and dense basalt aggregate form excellent road aggregate.

(iv) Deleterious materials. Materials which should not be used in concrete are listed above but it is necessary to consider their harmful effects in more detail.

The clay minerals montmorillonite and illite are commonly present in clays, shales, and tuffs; both absorb considerable amounts of water with consequent expansion and failure of concrete. Similarly, an aggregate with a high proportion of orthoclase or microcline (as in granite) may be unsuitable if the concrete is to be subject to intense heat (Krynine and Judd 1957). Both orthoclase and microcline have a high coefficient of expansion along one axis which can build up intense stresses in concrete where these minerals are abundant.

The expansive reaction between the alkali content of cement and aggregate containing opaline or chalcedonic quartz leading to gel formation is well known. Siliceous rocks, sedimentary rocks cemented with opaline silica, volcanic rocks, phyllites, and certain dolomitic and argillaceous limestones should all be avoided in concrete

171

aggregate (Gillott 1975); quartz is inert and can be safely used. In the British Isles it is possible to avoid reactive aggregate but in other areas of the world (e.g. Nebraska and parts of southern California) it is difficult to find a source of aggregate free of opaline silica (Krynine and Judd 1957). In such cases either low-alkali or pozzolanic cements are used, or an entraining chemical compound is introduced into the concrete; the latter increases the percentage of voids in the concrete and provides space for expansion. In general potentially reactive aggregates require careful testing before use.

Pyrite and marcasite are deleterious in concrete, especially in warm humid climates, where they oxidize, hydrate, and increase in volume. In concrete sewer pipes (Lenhart 1962) the presence of pyrite may lead to the generation of H_2S and the formation of sulphuric acid which attacks the CaO in the cement, forming fissures and cracks. Organic substances such as coal contain organic acids which inhibit the hydration of the cement and thus lead to a decrease in its strength and durability. Similarly, sulphates expand in concrete causing spalling and disintegration.

Clay coatings on aggregate particles must be removed by washing or they will prevent a bond forming between aggregate and cement; when clay-coated aggregates are used in roads this lack of bonding causes 'crawling' in the bitumen and break-up of the road surface. In cold climates, porous aggregates (sandstone, vesicular basalts, etc.) must be removed to eliminate 'freeze-thaw' effects.

In parts of the United States where many areas lack good-quality aggregate it may be necessary to upgrade the gravel (i.e. remove the deleterious constituents). This can be done either by vibrating screens (clastic fractionation) or by heavy media separation. The latter process separates rocks of different specific gravity (e.g. those of s.g. 2·60 from those of 2·45) and a separation of granitic rocks from limestones can be effected by immersing them in a dense liquid. At the Glen Canyon Dam (Lenhart 1962) the H.M.S. gravel was then used for facing concrete, terminal buildings, and other exposed concrete structures, and the normal washed gravel formed the bulk of the construction concrete.

In some sand and gravel pits the separation of heavy minerals may be a profitable sideline and in one pit in the United States the removal of gold largely contributes to plant running costs. In glass sands heavy minerals must be removed since the heavy minerals resist fusion and produce blemishes in the glass or may even destroy the furnace.

2.2. Occurrence of sands and gravels in Great Britain

With the exception of certain shield areas which have suffered severe glacial erosion and where weathering is slow, sand and gravel are ubiquitous. Subject to availability and quality most centres utilize the aggregate nearest at hand since this reduces costs of production. The British Isles has been fortunate in not only having significant resources of sand and gravel but in having them accessible to the large conurbations and expanding manufacturing and urban areas (Knill 1963*a*). However, conflicting demands for the use of land curtail unlimited expansion of sand and gravel extraction so that industry will be forced to expand its extraction of sand and gravel from offshore deposits or to develop alternative aggregate sources in the future unless more land is released for aggregate extraction (Verney 1976).

Unconsolidated sediments older than the Permo-Trias are not found in Great Britain although some Dalradian quartzites are easily crushed and are now worked for glass sand (Muckish, Co. Donegal, Eire). Broadly speaking, the bulk of British sand and gravel deposits are found either in relatively unconsolidated sands of Jurassic, Cretaceous, and Tertiary age, or in glacial and fluvial sands and gravels of Quaternary age. A loose yellow sand at the base of the Permian in County Durham is used locally as a building and concrete sand, and the Penrith sandstone of Cumberland is used as a moulding sand. Triassic sandstones form an exceedingly important source of moulding sands for non-ferruginous foundry work and also provide a resource of sands and gravels. Important moulding sands are found in the Triassic of the Lagan Valley in Northern Ireland, and these are exported to Scottish foundries and to the United States. While the Jurassic does not contain abundant sands, local deposits of sand such as the Bridport Sands and Yeovil Sands are used as building sands and the Kellaway beds in the Derwent and Humber Valleys are a source of refractory and moulding sands. The Cretaceous contains numerous sands of economic importance which are used as sources of building and glass sands, for refractory sandlime bricks, and as scouring sands. Tertiary sands occur in the south of England and form the principal economic sands and gravels of that area (Thanet sands, Erith sands, Woolwich and Reading Beds, Blackheath Beds, etc.). These sands have been utilized in a variety of ways according to their composition but are used to a very limited extent at the present.

The bulk of British sands and gravels are derived from Quaternary deposits occurring both on-shore and off shore. These deposits are divided into two types: those deposited outside the ice sheet (proglacial deposits) and those deposited within the limits of the ice sheet. The proglacial deposits are confined to the south of England, which is the only area of the British Isles not covered at some time by an ice sheet. They are mainly pebbly gravels and sands of fluvial origin occurring mainly in the Thames Valley. The rest of the British Isles was at various times covered by several ice sheets which, as they advanced and retreated, left a wealth of deposits in the form of moraines, glacio-fluvial deposits, and so forth. Meltwaters issuing from beneath the ice sheets deposited sands and gravels as eskers, kames, etc.; such deposits are worked extensively in the Midlands and the north of England. In Scotland glacial deposits are widespread although many of them are in remote and rather inaccessible areas. Aggregates in Scotland differ from those in England and Wales: they exhibit marked shrinkage when used in concrete. Such concrete can disintegrate after relatively short exposure to natural weathering. Most Scottish gravels are glacial or reworked glacial deposits and are very variable; their content can differ over very short distances. Aggregate from each pit must therefore be separately tested for potential shrinkage. Gravels from glacial deposits in the Southern Uplands which contain a high proportion of greywacke fragments yield the highest shrinkage value for Scottish gravels, presumably because of the clay and secondary minerals present in the rock matrix. Fluvial gravels in the Midland valley contain abundant coal fragments which are similarly deleterious in concrete.

Diatomite deposits of late glacial or postglacial origin are found at several localities in the British Isles, the most extensive deposits being found at Lough Neagh, Northern Ireland. Diatomite is important as a source of high-grade silica rock, being used in the manufacture of lightweight silica bricks and porous silica ware, as an abrasive, and as an absorbent in the manufacture of high explosives.

Deposits accumulating at the present time include beach sands and gravels, estuarine and river sands, and shingle. The gravels of many beaches, although marine in origin, may be re-worked from glacial or fluvial deposits off shore; in the south, such deposits are composed of well-rounded flint. Elsewhere in the British Isles, the material of gravel beaches is largely derived from glacial deposits and is of a variable nature. Shell sands form sources for agricultural lime; beaches consisting of cobbles and pebbles can be used as ballast, road aggregate, or in certain circumstances concrete aggregate. Wind-blown beach sands are also extracted and near Scunthorpe such blown sand is used in the laying out of pig-iron beds.

2.3. Utilization of sands and gravels

The bulk of the sand and gravel which is extracted is utilized by the construction industry, and a considerable proportion of it is used in roadmaking, either in a concrete or a bituminous base. Sand and gravel are also required in considerable quantities as a base course or fill where good drainage properties are required, e.g. in road sub-grades, drains, railway ballast, in foundation slabs under ware-houses to prevent humid conditions developing, and so forth.

Sands are extremely important in the manufacture of asphalt, mortar, plaster, floor screeds, filling mortars, foundry bricks, moulding sands, fillers, horticultural sands, terazzo flooring, optical and glass sand, water filtration plants, etc. The type of sand required will be governed by the purpose for which it is required. Sand of specific grade is also exported, even as far as Abu Dhabi, where it is used for the water filtration plant used in oil extraction. The economic importance of the sand and gravel resources and their utilization cannot be overestimated.

3. Road aggregates

Roads are vital in the establishment of communication and trade. Nevertheless, it was not until comparatively late in the history of man that roads, as opposed to well-defined beaten earth tracks, were developed. The first series of highways originated in Assyria in about 750 B.C. and were composed of either small blocks of rock or fired clay-bricks laid on an earth base. The blocks themselves were either closely packed, or fine clay or sand was packed between them.

Basically, the construction of Roman roads differed little from modern ones; indeed, cement of the quality found in Roman roads

TABLE 8.2
Construction layers of a modern road

Road course	Thickness	Important features	Major construction materials	
			Rigid pavement	Flexible pavement
Wearing course	13–38 mm	Aggregate attrition value and polished stone value are important in these layers	Concrete with granite aggregate	Rolled asphalt with bitumen coated chippings
Base course	38–76 mm		Air-entrained concrete	Hot-rolled asphalt
Road base	104–204 mm	Rock with high crushing strength which is also permeable to allow good drainage	Reinforced Concrete slab	Dense bitumen with aggregate
Sub-base	Variable	As for road base. Drainage layer of unbound aggregate	Crushed rock	Crushed rock
Subgrade	Rock base or soil, e.g. natural foundation	Natural foundations of clay, sands or unconsolidated gravels are compacted		

was not made again for many centuries after the fall of the Roman empire. The application of a bitumen dressing to the wearing course is also found in many Roman roads and it was this idea which was later adapted by MacAdam in the nineteenth century. After the Romans left Britain, their well-defined system of highways gradually decayed through lack of maintenance, and in the sixteenth and seventeenth centuries communications in Britain consisted of a series of rutted earth roads which during wet weather were often impassable. The Industrial Revolution, however, produced a need for the rapid transport of both raw materials and finished goods and saw the emergence of a well-constructed system of highways.

Modern roads are constructed to withstand heavy wear (Please *et al.* 1972). The layers of the pavement vary in thickness according to the volume of traffic the road is to carry.

The basic construction layers of a modern road are outlined in Table 8.2. In 1970 some 50 million tonnes of aggregate was used in Great Britain in the construction, maintenance, and improvement of roads and it is estimated that by 1980 30 million tonnes a year will be required for maintenance of existing roads and 58 million tonnes a year for new roads. These figures do not include the

Fig. 8.1. Crushing plant for granite, Turlough Hill, Eire.

quantities required for such schemes as a new London Airport or the Channel Tunnel, if these were to be constructed in the foreseeable future. In Great Britain, there is obviously no lack of natural resources but many of the sources are too far from the areas where they are needed and are uneconomical as transport costs become prohibitive. Other sources are also in areas of outstanding natural beauty, or lie within Forestry Commission or agricultural land, and their working is prohibited. Large resources of manufactured aggregate arise from industrial resources and offer considerable potential as road aggregate. They have the advantage in many cases of occurring in industrial areas, usually near to road-building centres, and transport costs should be cheap; their utilization is discussed in a later section.

The principal sources of road aggregate are in igneous rock or metamorphosed sediments (Fig. 8.1). In southern England, a large proportion of road aggregate is derived from the flint gravels of south-east England and the Thames Valley, and these are worked from Pleistocene and Recent sands and gravels. In Great Britain roadstones have been classified for many years according to the Trade Group Classification (B.S. 812, *Sources of Road Aggregate* 1968). This classification has several limitations when the engineering properties of the aggregate are concerned but obviously no simple classification will be entirely satisfactory. A petrological examination, coupled with the appropriate physical tests, is the only reliable means of assessing the properties of a particular roadstone. Hartley (1974) has proposed a simple classification of

TABLE 8.3

Classification of road aggregates
(After B.S. 812)

Basalt Group	Artificial Group
Granite Group	(All slags, fly ashes, etc.)
Gritstone Group	Limestone Group
Clastic Volcanic Group	Quartzite Group
Gabbro Group	Schist Group
Porphyry Group	Flint Group
Homfels Group	
(all thermally metamorphosed	
rocks except marble)	

road aggregates based on petrological characteristics; Table 8.3 outlines a proposed classification of road aggregate.

Ideally, a good aggregate for use in a road surface must be able to withstand abrasion and resist impact during wear. It must resist crushing during rolling, the fragments should be rough and angular, and they should be free of clay coatings or oxidized skins since these prevent the aggregate binding well in bitumen. In addition, the skid resistance of the aggregate in the wearing course should be high and absorption should be low so that the aggregate has a maximum resistance to frost. An unsound aggregate would wear rapidly, disintegrate when crushed mechanically, fail to bind in a bituminous base, absorb water and therefore be subject to disintegration through frost or chemical weathering, and have a low skid resistance. Naturally no single aggregate will fulfill all the requirements of a good aggregate but the type of road being built will limit the selection of many aggregates. Some aggregates, however (e.g. granite), while unsuitable for the wearing course are suitable for use in the road base or sub-base.

In order to determine the physical properties of road aggregate a series of physical tests has been devised: attrition test, abrasion test, impact test, crushing strength, absorption, specific gravity, density, and polished stone value (which measures the skid-resistance of the aggregate) (B.S. 812).

A petrological examination of the aggregate (Knill 1963b) will suggest how the rock will behave as a road aggregate. The main features which should be considered are: (a) the minerals present, (b) the proportions of different minerals, (c) the degree of mineral alteration, (d) the rock texture, (e) secondary minerals. Rocks composed of hard minerals such as olivine, plagioclase, quartz, epidote, etc. can be expected to wear better than rocks consisting of carbonates, micas, clay minerals, etc. Similarly, a rock with an interlocking texture or network of feldspar laths, such as a basalt, will be inherently stronger than a rock with a foliated texture which will have definite planes of weakness. An increase in grain size is partly accompanied by a decrease in strength, and a coarse-grained mosaic of crystals will not wear well since mineral grains and fragments of minerals will be plucked from the rock. Such a rock may, however, have a low skid resistance and could be used in the wearing course at dangerous junctions (Knill 1960). The polished stone value is also influenced by texture and mineral proportions but unfortunately those rocks which wear well as roadstones are also those which polish well under heavy traffic. The polished stone value of aggregates appears to be most closely related to the grain size of the rocks, the proportions of hard and soft minerals, and also the degree of weathering (Knill 1960; Hartley 1974). The cement present directly affects the usefulness of sedimentary rocks as road aggregates. Secondary cements composed of silica increase the strength and durability of the sediment; indeed, all sediments used in the British Isles as road aggregate have suffered secondary cementation by silica or carbonate. In other parts of the world, however, it may not be possible to avoid using poorly cemented sediments for aggregate. For a good-quality wearing course to be prepared, the road must then be artificially compacted or cemented artificially. The recrystallization of limestones also increases their strength and, provided that the grain size of the limestone is not too fine, an excellent road aggregate can be produced (e.g. the Carboniferous Limestone of the Pennines).

In those regions where sand and gravels are limited, crushed rock will provide concrete aggregates, the general specification for which will be similar, or identical, to that for concrete-making gravels.

A road aggregate which is extremely durable and yet has a high degree of skid resistance is difficult to find among natural rock aggregates. Such an aggregate may nevertheless be essential where skidding accidents are common or where the repeated closure of a road for repair is not only uneconomic but a nuisance. The development of synthetic aggregates is under investigation. So far, the most satisfactory artificial aggregate appears to be calcined bauxite laid in an epoxy-resin bitumen (Hartley 1974). The latter holds the aggregate firmly in position which ensures a good road texture. The aggregate is obtained by calcining bauxite at 1600 °C, which yields a welded vesicular texture composed of crystals of corundum and mullite. These minerals are exceptionally hard and therefore extremely durable. In order to reduce crushing and impact the calcined bauxite fragments are small (1·2 and 2·8 mm). Unfortunately, this aggregate is expensive and its use can be justified only at localities with a very high accident record; cheaper road aggregate with a high skid-resistance may be provided by other sources such as pulverized fly-ash, metallurgical slags, devitrified glasses, etc.

4. Large-scale construction stone

Large-scale aggregate or rubble is durable, irregular broken stone used in the construction of jetties, breakwaters, rockfill dams, road embankments, and various engineering construction works. The blocks range from small fragments (4 mm) to blocks weighing up to 10 tonnes or more. The rubble is either obtained from quarries which produce large-scale rubble, or it is obtained as a by-product from quarries which produce building stone or crushed stone. Naturally occurring rubble, such as that found in boulder clay, is sometimes used but the cost of removing fines such as clay is usually prohibitively expensive and the rock boulders may also be too variable in size and rock-type for the deposit to be economic. Obviously, for the production of large blocks the rock must be massive with all joint systems widely spaced. Where igneous rocks are being considered, the texture should be interlocking, for this will give a stronger rock. Sedimentary rocks should be well cemented or recrystallized. As much of the large-scale aggregate is used in construction where the rubble will be in contact with water, soluble rocks are avoided as are shales, since they often, on wetting, decompose to mud or clay and may totally disintegrate (Penman 1971). Porous or lightweight rock is also unsuitable, since rocks of this type are usually weak. In practice, a blasting trial will often reveal the suitability of a rock; any rock which shatters on blasting or produces a large percentage of fines should be rejected. The most suitable sources of large aggregate are unweathered granites, basalts, gritstones, massive sandstones, and recrystallized limestones.

Other factors which will influence the selection of a quarry for large aggregate are its distance from the site where it is required, access to public roads, the thickness of overburden which must be removed during extraction, and the availability of labour at the quarry site. In remote areas these factors when costed may be equally as important as the suitability of the aggregate. The various types of constructional stone will now be considered.

Riprap. Riprap is rubble that is used to protect the upstream face of a dam from wave erosion. It is also used in the construction of road embankments, levees, in railway ballast, and at the base of sea cliffs to prevent erosion by surface run-off. Normally rip-rap ranges in size from 4 mm to large blocks weighing up to 1 tonne or more. The dimensions of the rubble for embankment dams are governed by the estimated wave size. For example for wave heights of 2 to 10 ft a blanket of riprap 12–36 inches thick would be laid with the sizes of the riprap between 0·25 inch and 400 lb weight (Treasher 1964).

Rockfill. Anyone who has spent half an hour damming a small stream draining a beach will quickly have found that a more effective dam can be obtained by using relatively large blocks packed with graded material than by using sand alone. Loosely dumped rockfill of this type was used during the California gold rush of 1850 where earth fill was absent. Modern dams require a rather more sophisticated approach than either a 'beach dam' or a temporary affair to assist prospectors. Stone for rockfill dams may contain blocks up to 10 tonnes in weight and the fill must drain freely to prevent the build-up of pore pressures (Fig. 8.2). The amount of clay or fines is usually limited, although graded aggregate is used for interstitial fill. Settlement due to load, or reorientation of rock fragments when saturated, is common in rockfill embankments. The introduction of graded material lowers

Fig. 8.2. Coarse rockfill, composed of gneiss, in contact with filter layer, Nyumba ya Mungu Dam, Tanzania.

the number of voids present and therefore reduces the amount of settlement (Penman 1971). Mechanical compaction also hastens settlement. Compacted rockfill dams are the highest in the world and include Nurik (U.S.S.R., 285 m), Scammonden (Great Britain, 70 m), and Llyn Brianne (Great Britain, 91 m). Post-construction settlement of rockfill dams is important, especially where a rigid structure such as a bridge or road either abuts the embankment or utilizes the embankment. The M 62 near Huddersfield is carried along the crest of the Scammonden Dam and differential settlement between the embankment and the motorway approaches must obviously be at a minimum. Rockfill is often preferred to earthfill in the construction of road embankments. The embankment slopes can be steeper than with earthfill and therefore a smaller volume of material is required. The embankment can also be raised and compacted more quickly since pore pressures are virtually absent in rockfill.

Derrick and armouring stone. In the environment of break-waters, dams, and coastal defence works there are specific areas where wave velocity can be expected to be high and scouring can occur. In such situations very large blocks weighing between 0·5 and 20 tonnes, requiring a crane or derrick to emplace them, are called for. Rectangular blocks are preferred. The stone must obviously be dense, hard, and resistant to abrasion and the blocks must be large enough to prevent waves moving them and using them as erosive agents. The construction of jetties and breakwaters either as coastal or harbour protections requires an enormous amount and weight of rock since these structures are built where erosion is especially severe. The Crescent City breakwater in north-west California, which is 4700 ft long, required over 1 million tonnes of armouring stone. The impact of heavy seas has been known to shift rocks weighing 100 tonnes over distances exceeding 20 ft to the top of a breakwater. Breakwaters normally are constructed in three or four zones, each of which contains rock of a particular grade, the larger rock forming an outer, protective layer. In some instances, large-scale aggregate may not be available, or may be uneconomic, and so large blocks of concrete are used instead. These are generally less satisfactory, for concrete blocks tend to settle more and leave only a small percentage of voids which fail to disperse wave energy. Shaped concrete blocks (tetrapods, quadripods) weighing up to 25 tonnes have been designed and are in use as a top armouring in breakwaters, for example along the California coast (Santa Cruz and Crescent City) and in French Mediterranean ports.

5. Aggregate derived from industrial waste and its utilization

Aggregate derived from a variety of industrial processes forms a considerable bulk of 'waste' material (Gutt *et al.* 1974). Some industrial residues are fully utilized (e.g. copper slags) whereas others have merely been dumped with consequent dereliction and loss of valuable land for either building or agriculture (Barr 1969). The utilization of industrial aggregate is not only a wise investment on the part of the plant producing it but it avoids the despoilation of land by waste heaps; and the substitution of industrial aggregates for natural products reduces the further depletion of natural resources. The uses for industrial aggregate are numerous although some aggregates are restricted in their application by their chemical composition, shrinkage rates, and so forth.

5.1. Man-made industrial aggregate

Iron and steel slags. Blast-furnace slag (BFS) is produced in considerable quantities during iron and steel production and is drawn off the top of the melt as a froth. It consists mainly of

TABLE 8.4
Industrial use of iron slag

Type of slag	Use	Percentage used
Air-cooled	Roadstone (mainly as coated chippings) Railway ballast Concrete aggregate Filter media Dry slag	95·4, of which 79·4 is used as roadstone
Foamed and expanded slag	Brickmaking Roof and floor screens Structural reinforced concrete	31
Granulated slag	Cement (an activator such as Portland cement or gypsum + Portland cement must be added to enable the slag to set)	1·3

silicates and alumino-silicates of calcium and other bases. By selective cooling processes three different types of slags are produced: (a) air-cooled, (b) foamed, and (c) granulated slag. Air-cooled slag cools slowly permitting crystal growth whereas foamed slag is sprayed with intense jets of water to produce a foamed mass not unlike pumice in appearance; granulated slag is cooled rapidly to produce a glass. Almost all iron slag is processed for industrial use as summarized in Table 8·4.

Steel slags resulting from the fusion of pig iron with limestone or dolomite are high in calc-silicate minerals. After a period of weathering for the hydration of excess calcium they produce a good wearing roadstone with satisfactory resistance to polishing; steel slags are also used as fill.

Waste from coal-fired power stations (pulverized fly ash, PFA, and furnace bottom ash FBA). PFA and FBA are produced as waste in considerable quantity from coal-burning power stations. Of the 187 power stations in England and Wales, 136 are coal-fired (Gutt *et al.* 1974); 75 of these are fired with pulverized coal and produce PFA; the remainder use lump coal and produce FBA or clinker. In 1971–2 coal-burning power stations in Britain produced 9·8 million tonnes of PFA and 2·3 million tonnes of clinker. The Central Electricity Generating Board has set up an Ash Market Department to investigate the disposal and utilization of residue from power stations. PFA is disposed of in various ways: through settlement in lagoons adjacent to power stations which are eventually grassed over, by storage in heaps, as fill for clay pits; or dumped at sea outside the 7-mile limit and at depths greater than 20 fathoms; dumping at sea is practised only in the north-east. The principal industrial applications of PFA are in road fill, in embankments requiring low-density material, as fill for building sites, and as lightweight aggregate in aerated concrete blocks. A certain amount is used as an additive in cement and concrete, in fired clay bricks, and as industrial filler.

Metallurgical slags. Metallurgical slags are produced as a consequence of the melting of zinc, lead, copper, and tin during metallurgical processes. Lead smelting at Avonmouth alone produces 70 000 tonnes a year of lead–zinc slags, the bulk of which is disposed of as bulk fill, as aggregate for footpath construction where high strength is not required, and as a constituent of calc-silicate bricks. Other possible uses at present under investigation are in asphalt and concrete, as a medium for water filtration, and as heat-retentive blocks in night storage heaters. Lead–zinc slags can also be used where dense aggregates are required, as in radiation shields or noise insulators. Slags from copper smelting are completely utilized in grit blasting. Tin slag is produced at two smelting works at Liverpool and Hull; the resulting slag is glassy and chemically stable. Its use as a roadstone is limited by low skid resistance and it is restricted to the sub-base and base coarses; similarly, in concrete making it is restricted to low-strength concrete.

Chemical by-product aggregates. Calcium sulphate is produced as a by-product in several chemical process, the principal being the manufacture of phosphoric acid where it is known as phosphogypsum. For every ton of phosphoric acid produced, 5 tons of gypsum are produced as well, and this constitutes a valuable source of gypsum and possible sulphur. Both phosphate and fluoride must be removed from industrial gypsum before it can be substituted for natural gypsum and anhydrite in the manufacture of plaster and plaster board.

Waste glass. It is estimated that every year 115 000 tonnes of glass bottles and jars are thrown away in refuse dumps throughout Britain. Some container glass which is rejected is recycled within the glass industry, either being re-used or crushed for use as cullet; some is used in the production of glass fibre. Some European glass manufacturers have been able to use up to 40 per cent cullet in glass production but in the United Kingdom the proportion is usually 12 per cent. The main problem with waste glass lies in sorting, either by the housewife or by the local refuse authorities. Small amounts of detritus, such as a bottle cap, can completely ruin a furnace and so far the collecting and sorting of glass in Great Britain appears to be too expensive although it is adopted in other countries. In the United States research has gone into the recycling of waste glass in the production of bricks, insulating wool, and lightweight aggregate. Glass has also been used in the U.S.A. to form 'glasphalte', an asphalt in which glass replaces limestone. Unfortunately as a wearing course it has a low skid resistance and its use is limited. Glass used in concrete may react with the cement, which can limit its application in that field.

5.2. *Natural aggregate produced by the extractive industries*

Colliery spoil. Half of the material removed during coal production during mining is left as spoil and the mechanization of mines has considerably increased the amount of unwanted materials. Spoil comprises the waste sedimentary material remaining after coal extraction; it includes sandstone, mudstone, siltstone, shale, seat-earths, limestone, etc. Normally the spoil is tipped and at present approximately 2000 million tonnes are dumped annually over 10 692 ha. The spoil is dumped as either burnt or unburnt colliery spoil. Unburnt colliery spoil, because of its organic content, is liable to spontaneous combustion and was held in disfavour although burnt colliery spoil found a ready outlet. It has, however, been shown that if the spoil is well compacted the problem does not arise. In the Thorn by-pass (M 19), 1·25 m tonnes of unburnt colliery spoil were supplied by the near-by Hatfield Colliery; colliery spoil has also been used on the Liverpool outer ring road and on the M 62. Altogether, 7 million tonnes a year are used in road embankments, in reservoir beds, and as fill on building sites; other uses include brickmaking, cement, and lightweight aggregates. The high sulphate content, especially in burnt colliery spoil, produces sulphuric acid, which in turn attacks the cement; so its use in concrete is restricted. Because of the availability of good quality brickmaking clays the use of colliery spoil in brick-making has considerably decreased. Lightweight aggregate can be produced from colliery spoil and used in concrete requiring low strength. Roadstone with a low skid resistance can also be produced by heating colliery spoil with calcined bauxite. Colliery spoil can also be used in the production of Portland cement, where it can replace the clay materials required to form cement.

China clay waste. In Britain, china clay is extracted for the ceramic and paper industry from the Armorican granites of southwest England. The china clay occurs as kaolin which has formed from the decomposition of orthoclase. The rotted granite is further broken down by spraying the pit sides with powerful jets of water and the slurry is pumped through two separation plants (Gutt 1974). An enormous bulk of residue is produced, which contains not only kaolin but also quartz, mica, waste rock, and overburden. It is estimated that the production of 1 tonne of china clay produces 9 tonnes of waste, containing 3·7 tonnes of coarse sand (mainly quartz), 2 tonnes of waste rock, 2 tonnes of overburden, and 0·9 tonnes of micaceous residue. About 22 million tonnes of this waste is dumped each year, the coarse aggregates in the familiar conical mountains and the micaceous residue in mica dams where settlement of the phyllosilicates occurs. At present about a million tonnes is used as bulk fill for building material and a certain amount of the waste is screened to supply four grades of sand for local needs. At Rocks Pit in the St. Austell area, concrete blocks are produced and at Lee Moor, calcium silicate bricks are made; the local market absorbs the total production of these by-products. China clay sands are especially suitable for the manufacture of white concrete blocks and the prospects for this market appear to be good. China clay can also be screened for use as plastering and mortaring sands. Coarse aggregate (mainly waste rock) can be separated and crushed for use as fill and road aggregate although its grading does not fulfil the specifications of B.S. 594 and it is therefore suitable only as a sub-base aggregate. A recent investigation into the substitution of micaceous aggregate for asbestos in floor tiles appears to be promising. The micaceous residue is also mixed with wastes from paper mills to produce a type of board having properties similar to asbestos sheeting and plasterboard. Consideration has been given to the transport of sand from south-west England to London for construction purposes; but at a present cost of £3–4 per tonne for transportation, this appears unlikely.

Waste from slate quarries. The principal areas of slate production in the British Isles occur in north Wales and the Lake District; smaller amounts are produced in Cornwall and south Wales. Slates are metamorphic argillaceous rocks (or fine-grained volcanic tuffs) characterized by a well-developed cleavage. In a good-quality slate the cleavage planes will be smooth and regularly spaced, the composition being uniform; accessory minerals, such as calcite, epidote, and pyrite, will be present only in minute amounts. Slate is produced either by quarrying or mining, and is then split into roofing slates or slabs for industrial or architectural purposes. Its extraction results in the production of enormous amounts of waste: the ratio of unusable rock to waste is about 1:20. The bulk of the waste is tipped although a certain amount is stored underground in disused mine channels. In parts of north Wales where the land is agriculturally unproductive, tipping has met with little opposition although it is unsightly and in the vicinity of National Trust land or National Parks slate refuse tips reduce the amenity value of the area. In proportion to the bulk of slate waste produced, little use is made of slate aggregate at the present time. Some is ground and provides an inert filler for bitumen, linoleum, paper, and rubber; a small amount is used as road aggregate. Some slate aggregates have been expanded by heat treatment to produce filter material or lightweight aggregate. Current production of expanded slate aggregate in the U.S.A. exceeds 630 000 tonnes. The production of slate-lime bricks is currently under investigation in Britain but, although bricks of adequate strength could be produced by autoclaving limestone and crushed slate, the bricks have a high drying shrinkage and are susceptible to frost damage in severe conditions.

Quarry waste. Quarry waste includes poor-quality stone and varying amounts of topsoil, overburden, and quarry fines which fall below the specifications for rockfill or road aggregate. The disposal of topsoil presents no real problem but the remaining wastage after screening can be sold either as aggregate for cement-stabilized base or sub-base in highways. Other outlets include brickmaking and, for suitable waste, the manufacture of concrete bricks. The principal deterrent to the utilization of quarry waste generally lies in the expense of disposal.

6. Building stones

6.1. Bricks
The earliest building stones used by man were loose stones and boulders which were initially handpacked (as in dry stone walls) and at a later date were packed with a clay mortar. The earliest walls of Jericho (*c.* 8000 B.C.) were of this type. Somewhat later walls (7000 B.C.) were mortared and covered with plaster. Mud bricks have been found in structures dating back to 4800 B.C. but were probably in use much earlier. At a later date, mud bricks were fired and a brick with a much greater strength and durability was developed. When fired bricks were packed with cement mortar they could be used to construct fairly elaborate buildings. The Ziggurat of Ur (2100 B.C.) was built to a height of 70 ft from mud bricks and faced with fired bricks set in a natural bitumen cement.

Aggregates, sand, gravel, and constructional stone

The Romans were experts in brickmaking and also produced an excellent cement mortar. After the withdrawal of the Romans most European buildings were constructed in timber or wattle faced with plaster. During the Middle Ages, the production of hand-made bricks for prestige establishments (royal palaces, etc.) was a craft industry and the hand-made bricks of England were in demand throughout Europe. The fires which ravaged the wooden cities of Europe led to an increase in brickmaking; after the Fire of London the bulk of rebuilding was carried out in brick or natural stone. By the time of the Industrial Revolution, not only had large areas of Europe been deforested but the developing industrial centres led to an unprecedented demand for cheap and durable building materials. Brickworks sprang up wherever a workable clay deposit occurred near an industrial centre, and, with the development of machinery to cut and shape standard bricks, brickmaking became a major industry in its own right.

In 1971, the brick industry of Great Britain produced 6·5 thousand million bricks, principally fletton bricks from the Oxford Clay, which accounts for 43 per cent of the United Kingdom brick production (Open University 1974). Within the British Isles, bricks are made from clays of Carboniferous and post-Carboniferous age, in particular the London Clay, Gault Clay, Weald Clay, Lower Lias, Oxford Clay, Keuper Marl, and Etruria marl. Recently a new brickworks has been opened up at Avonmouth where, using alluvial clay as the medium, new methods of brickmaking are being investigated. The production of bricks fluctuates with the housing market. A normal three-bedroom detached house requires 7000 facing and 4500 common bricks in its construction.

The requirements for brick are that it should be durable, strong, of low porosity, and of attractive appearance (Open University 1974). Bricks are normally divided into three types: (1) 'commons', for general building, (2) facing bricks, with an attractive appearance, and (3) engineering bricks, special bricks with a high strength and low absorption.

Although many clays are used for brickmaking they usually have the following properties in common.

(i) The water content lies between 17·5 and 19 per cent. While the clay must be plastic and workable, too much water can cause irregular shrinkage when the bricks are cut and dried. The final cost of brick manufacture can be greatly increased if too much water has to be driven off during firing.

(ii) The clay mineral content of the clay should not be greatly in excess of 20 per cent: too high a clay mineral content produces too plastic a clay. A pure kaolinite clay is not plastic; montmorillonite is too absorbent and shrinkage rates will be high. Generally clays for brickmaking contain irregularly crystallized kaolinite and illite which will take up sufficient moisture to make them plastic but are not too absorbent.

(iii) A minimum of 5 per cent of calcium carbonate should be present, since this prevents shrinkage of the brick during firing. If the proportion of calcium carbonate exceeds 15 per cent then carbon dioxide is expelled and the brick becomes porous and weak and its frost resistance is poor.

(iv) A small amount of carbonaceous matter (usually less than 5 per cent) is desirable as this gives off heat during firing and reduces the actual furnace costs. The brick clay which fulfils all these requirements to the optimum extent is the Oxford Clay, and the majority of brickworks in Great Britain are situated on its outcrop.

Refractory bricks. Although pure kaolinite clays are not suitable for building bricks, kaolinite clays are important in the manufacture of refractory bricks for lining purposes. Kaolinite has a melting point of 1800 °C and does not contain fluxes which lower the furnace temperature. Kaolinite is derived from china clay in Cornwall and from the Bovey Tracy Beds of Dorset.

Building stones. Rock was not quarried extensively as building stone until the discovery of metals in about 3000 B.C., although flint implements were used earlier to trim loose rock blocks. One of the best early examples of the quarrying of rectangular blocks are the limestone blocks which were used in the construction of the Pyramids in 2800 B.C. Stonework became increasingly important in the construction of defensive and religious buildings comparatively early in history and, although the general tendency was to use local stone where possible, a considerable trade was built up in the more ornamental stones. Today, for economic reasons, few buildings are built entirely in dressed stone. The walls are no longer load-bearing stone walls but the load is taken by a steel frame and the shell is composed of concrete blocks, either pre-cast or cast *in situ*. These may be faced with a veneer of natural stone which may be no greater than 5 to 15 cm thick. In selecting a building stone the following features are of importance (Blyth and de Freitas 1974): strength, durability, spacing of joints, and appearance of the stone when dressed. Economic features which must be considered are: ease of quarrying, thickness of overburden, distance of the quarry from the site where the stone is required, and drainage of the quarry.

Any building carries a load which in simplest terms is dependent on its height, base area, and the material from which it is built. The average strength of various typical rocks is given in Table 8.5. Generally speaking, igneous rocks, some quartzites, and some sandstones have the highest compressive strengths. Fresh unweathered basalts may reach a strength of 400 MN/m² (Krynine and Judd 1957); typical figures are given in Table 8.5.

TABLE 8.5
Compressive strengths of various rocks

Rock-type	Compressive strength (MN/m²)
Basalts, diabase, some quartzites	250
Fine grained granites, diorite, basalt, well-cemented sandstones, limestone, quartzite	160–250
Sandstone, limestone, medium-and-coarse-grained granites, granodiorite	60–160
Porous sandstone, limestone, mudstone	30–60
Tuffs, chalk, very porous sandstone/siltstone, shales	< 30

The strength of a rock is influenced by the texture. A fine-grained igneous rock with an interlocking texture will be stronger than another with a coarse-grained hypidiomorphic-granular texture. The controlling factor in sedimentary rocks is the nature of the cement; the highest compressive strength in sediments occurs when the cement is siliceous and will be lowest in sediments cemented with clay, or ferruginous material. The compressive strength is also influenced by the direction of bedding or foliation and is greatest normal to the bedding. For this reason. if none other, sedimentary blocks should always be laid with the bedding parallel to the horizontal.

The saturation of a rock also decreases its strength and this in its turn is directly related to its porosity. The higher the porosity of a rock, the lower will be its strength and its resistance to frost action. It is, however, important to distinguish between the potentially available voids in a rock (porosity) and the extent to which those voids can become saturated (absorption).

The building stone used should be reasonably resistant to weathering, which in excess can destroy not only the beauty of a building stone but can lead to its failure. Industrial processes release large amounts of carbon dioxide and sulphur dioxide into the air; these fall in rain as carbonic and sulphuric acids which can have serious effects on limestone buildings. These weak acids slowly dissolve the limestones in time and gypsum is also produced. The gypsum crystals produce volume changes within the rock and form a disfiguring sulphate skin (Schaffer 1972).

Porous sandstones, travertines, etc. are susceptible to frost damage and in climates where frost is common are generally unsuitable. For example, the travertines of the Balearic Isles and Spain are admirably suitable to a Mediterranean climate which receives little frost but are totally unsuitable, except for interior work, north of the Mediterranean. A similar example can be given from the very attractive but open-textured limestones of Florida, which are suitable for building stone only within the southern states. The durability of sediments is also affected by the bedding and, if the bedding-plane is laid vertically in exterior work, flaking parallel to the bedding planes will occur.

The presence of secondary minerals such as pyrite, chlorites (e.g. phrenite), serpentines, etc. can produce discoloration of the building stone and may even cause failure through volume expansion.

For ease of working, and in order to obtain blocks of maximum size, joints and bedding-plates should be widely spaced. Widely spaced joints also reduce the amount of quarry waste and, therefore, the cost of the stone. Furthermore, when joints are widely spaced, the rock should be less weathered, although this is very dependent on the climate and on circulation of groundwater.

Although strength and durability are undoubtedly the most important features when selecting a building stone, the appearance of the dressed stone must also be considered. The attractiveness of many English villages is considerably enhanced by the use of local building stone, especially the various cream and yellow Jurassic limestones. Today, many concrete buildings are faced with attractive British and foreign building stone such as laurvikite from Norway which is exported all over the world and owes its demand to the attractive blue of silvery schiller of the feldspar. Granite facing slabs are normally 2 in thick. They give a good, durable facing to a building, for granite has a low porosity and high resistance to frost.

Virtually every country possesses attractive building stones which will serve a miscellany of purposes whether for facing, large aggregate, rock fill, or road aggregate. Even though modern architecture generally utilizes concrete blocks and favours a simpler and severer finish than buildings in the past, there is still a high demand for high-quality building stone.

7. Conclusion

It has been recognized (Verney 1976) that the present pattern of the British sand, gravel, and aggregate industry must remain in existence for some years to come. Nevertheless, it is likely that changes will be in hand by the 1990s in view of the working-out of all the gravel-bearing land of south-east Britain. For the future, there are interesting possibilities: the increased use of dredging, mining aggregates underground, the creation of super-quarries in granite masses, and the expansion of waste processing. An overriding issue will, however, remain the question of transport, which may well become the prime factor in the cost of aggregates extracted in the next century.

References

BARR, J. (1969). *Derelict Britain*. Penguin Books, London.

BLYTH, F. G. H. and DE FREITAS, M. H. (1974). *A geology for engineers.* (6th edn.), Edward Arnold, London.

British Standards Institution. B.S. 410: 1943. *Test sieves.*

—— B.S. 812: 1953. *Sampling and testing mineral aggregates, sands and fillers.*

—— B.S. 882, 1201: 1954. *Concrete aggregates from natural sources.*

—— B.S. 1198, 1199, 1200: 1955. *Building sands from natural sources.*

—— B.S. 1232: 1945. *Natural stone for building.*

GILLIOT, J. E. (1975). Alkali-aggregate reactions in concrete. *Eng. Geol.* **9**, 303–26.

GUTT, W., NIXON, P. J., SMITH, M. A., HARRISON W. H., and RUSSELL· A. D.

(1974). A survey of the locations, disposal and prospective uses of the major industrial by-products and waste materials. *Build. Res. Curr. Pap.* CP 19/74.

HARTLEY, A. (1974). A review of the geological factors influencing the mechanical properties of road surface aggregates. *Q. J. Eng. Geol.* **7**, 69–100.

H.M.S.O. (1968). *Sources of road aggregate.*

KNILL, D. C. (1960). Petrographical aspects of the polishing of natural roadstones. *J. appl. Chem., Lond.* **10**, 28–35.

—— (1963a). *The geology of the sand and gravel deposits of Great Britain, their deposition, distribution and utilization.* Sand & Gravel Association.

—— (1963b). The value and application of petrology to the production of natural aggregate and building stone. *Quarry Mgr's J.* **47**, 27–30.

KRYNINE, D. P. and JUDD, W. R. (1957). *Principles of engineering geology and geotechnics.* McGraw-Hill, New York.

LENHART, W. B. (1962). Sand and gravel. *Rev. Eng. Geol.* **1**, 187–96.

The Open University. (1974). *The Earth's Physical Resources, 4, Constructional and other Bulk Materials.*

PENMAN, A. D. M. (1971). Rockfill. *Symp. Rock Mech. in Highway Constr. Newcastle upon Tyne.*

PLEASE, A. C. and PIKE, D. C. (1972). *The demand for road aggregates.* Road Research Laboratory, H.M.S.O.

Sand & Gravel Association. (1967). *Pit and quarry textbook.* MacDonald, London.

SCHAFFER, R. J. (1972). Weathering of natural building stones. *Spec. Rep. natn. Bldg. Stn. D.S.I.R.* **18**.

TREASHER, R. C. (1964). Geologic investigations for sources of large rubble. *Engng. Geol. Case Hist.* **5**, 31–44.

VERNEY, R. B. (1976). *Aggregates: The way ahead.* H.M.S.O.

9. Geology and the cement industry

1. Introduction

The world cement industry currently consumes about 1000 million tonnes of basic raw material annually; it can fairly be described as a major user of the earth's natural resources. In 1974 the cement output of the United Kingdom was about 18 million tonnes, requiring 30 million tonnes of basic raw materials, exclusive of inherent moisture, for its manufacture; of this, 25 million tonnes were limestone or chalk and the remaining 5 million tonnes were mainly argillaceous materials such as shale or clay. Additionally, some four million tonnes of coal, or its equivalent in oil or natural gas, were consumed in the manufacturing process, and about a million tonnes of gypsum were also utilized. The quantity of water used by the United Kingdom industry in 1974 was in the region of 10^9 litres.

In order to comprehend the role of geology and the geologist in the cement industry it is necessary to understand the nature of the product and the process of manufacture, since the raw materials are used in an essentially untreated form to create a strictly controlled mixture of chemical compounds, the characteristics of which are largely dependent on the selection and blending of the raw materials. As minimal variation in the quality of the product can be allowed, suitable materials must always be available to produce, after blending, a raw mix that is essentially constant.

The *Oxford English Dictionary* defines cement as 'any powdered substance that, made plastic with water, is used in a soft and pasty state (which hardens on drying) to bind together bricks, stones, etc., in building, to cover floors, walls, etc., or (with a suitable aggregate) to form concrete'. Thus the term includes a range of substances from limes, through hydraulic limes and natural cement to Portland Cement. The chief difference between the manufacture of Portland Cement and the limes is the difference in temperature to which the raw materials are heated during production. The limes are prepared by heating pure limestone in the case of fat limes, or argillaceous limestone for hydraulic lime or natural cement, to a calcining temperature, whereas Portland Cement is, in the words of the British Standard specification (B.S. 12), 'manufactured by intimately mixing together calcareous or other lime-bearing material with, if required, argillaceous and/or other silica, alumina or iron-oxide bearing materials, burning them at a clinkering temperature and grinding the resulting clinker so as to produce a cement'. Numerous types of Portland cement have been developed over the years for various purposes, all being essentially modifications of ordinary Portland cement.

Some naturally occuring and artificial materials react with lime and water to form compounds with cementitous properties; these materials are known as pozzolans. The original pozzolans consisted of certain volcanic tuffs, although materials with pozzolanic properties can be prepared by burning certain clays and shales at appropriate temperatures. Pozzolanic cements are prepared by grinding together pozzolan and either Ordinary Portland Cement clinker or hydrated lime.

Complete details of the various cement types are given in Lea (1970) as well as in the appropriate British Standards and the American A.S.T.M. publications.

Although about twenty different types of cement are manufactured in the United Kingdom, more than 80 per cent of the total production is in the form of Ordinary Portland Cement (O.P.C.), and it is therefore this product which will be discussed in detail.

2. Chemistry of Portland cement

When the analysis of a typical Portland cement clinker is represented in oxide form, the oxides CaO, SiO_2, and Al_2O_3 commonly constitute some 93–4 per cent of the material. If to these is added Fe_2O_3, the four oxides can represent up to 98 per cent or more of the total chemical content. During its manufacture cement clinker is heated to a temperature at which sintering occurs and, although complete melting of the components does not take place, adequate combination is attained during manufacture. The resultant product, the clinker, is very similar in many ways to an igneous rock, being composed of a number of minerals. The relationships of these minerals to each other and to naturally occurring minerals is clearly shown by the ternary system $CaO-Al_2O_3-SiO_2$, given in a simplified form in Fig. 9.1. Although the quaternary system $CaO-Al_2O_3-SiO_2-Fe_2O_3$ more accurately represents the composition of Portland cement, the effect of the iron can be considered minimal for the purposes of this brief description. The field of the Portland cement composition is shown on the diagram and can be seen to extend into the fields of tricalcium silicate and calcium orthosilicate.

The major constituent in Portland Cement clinker is tricalcium silicate, variously written Ca_3SiO_5, $3CaO.SiO_2$, or commonly abbreviated to C_3S when, using the cement chemist's notation, $CaO \equiv C$, $SiO_2 \equiv S$, $Al_2O_3 \equiv A$ and $Fe_2O_3 \equiv F$. It exists in a number of polymorphic forms, three of which are triclinic and two are monoclinic, and a high-temperature form which is rhombohedral. In cement clinker it is modified by alumina and magnesia in solid solution and is known as alite, which normally refers to the monoclinic form.

Calcium orthosilicate, Ca_2SiO_4 or $2CaO.SiO_4$ (C_2S in the cement chemist's notation), also has a number of polymorphs of which the β phase normally occurs in Portland cement clinker. The stabilization of the β form is due to the presence of P_2O_5, alkalis, or Al_2O_3 in solid solution which inhibit the inversion to the stable low-temperature, but non-hydraulic, γ form. The impure β form of calcium orthosilicate or dicalcium silicate is known as belite. As well as calcium silicates, aluminates are present in cement clinker; the most common is tricalcium aluminate, $Ca_3Al_2O_6$ or $3CaO.Al_2O_3$, which in the industry is commonly abbreviated to C_3A. The main compound containing the iron occurs as the indefinite mineral brownmillerite, which is essentially a solid solution between dicalcium ferrite, $Ca_2Fe_2O_5$, and a hypothetical compound dicalcium aluminate, $Ca_2Al_2O_5$; it commonly approximates to an equimolecular combination of these, and is referred to as tetracalcium alumino-ferrite or abbreviated to 'ferrite phase' and written C_4AF.

The four compounds mentioned, alite, belite, tricalcium aluminate, and ferrite phase, all form hydrates when ground and mixed with water; the two silicates make the major contribution to concrete strength development. The setting time of cement is controlled by adding gypsum to the cement clinker when it is ground to produce the final cement powder; the effect of gypsum is

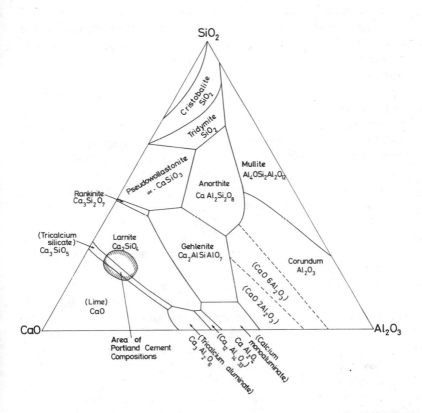

Fig. 9.1. Ternary diagram for the system CaO–Al$_2$O$_3$–SiO$_2$. Minerals not occurring naturally are shown in parentheses. Calcium orthosilicate, Ca$_2$SiO$_4$, occurs in four forms: α–Ca$_2$SiO$_4$, not naturally occurring; α′–Ca$_2$SiO$_4$, bredigite, not naturally occurring; β–Ca$_2$SiO$_4$, larnite, found in nature; γ–Ca$_2$SiO$_4$, lime-olivine or shannonite, reported as naturally occurring. Lime, although not normally occurring in nature has been reported from Vesuvius. Based on Rankin and Wright (1915) with corrections by Bowen and Greig (1924) and by Lea (1970).

to modify the hydration of the C_3A, which would otherwise tend to cause quick setting.

3. Manufacture of Portland cement

The nature of the raw materials strongly influences the type of manufacturing process chosen for a green-field site; if new raw materials are to be used in an existing plant, the method of manufacture at that plant will have a direct bearing on the choice of raw materials. It is therefore important for a cement geologist to understand the various ways in which raw materials can be prepared and processed so as to produce cement clinker.

The aim of the various manufacturing processes is to combine the compounds contained in the raw materials to form the silicates and aluminates of the cement clinker. Combination can be achieved in a raw mix prepared from pure calcium carbonate and silica by the use of multiple sintering techniques and very high temperatures; alumina and ferric oxide act as fluxes enabling the reaction to take place at a commercially economic temperature. At the temperature at which the raw materials are heated, only about 20–30 per cent of liquid melt is formed, the chemical reactions taking place partially through the liquid flux and partially in the solid state and therefore fairly slowly. The end-product of the manufacturing process depends not only on the overall chemical composition of the mixture of raw materials making up the feed to the process, but also on the physical nature of the raw materials; other important factors are the size of the individual grains in the feed material and the length of time during which heating takes place.

In the great majority of modern cement works, the heating or burning process takes place in a rotary kiln. This consists of a cylindrical steel shell lined with refractory bricks which is slightly inclined towards the end at which firing takes place. The fuel is powdered coal, oil, or gas injected into the kiln by means of an axially placed pipe. The kiln rotates at about one to two revolutions per minute and the raw material feed, passed into the kiln at the higher or 'cold' end, slowly travels through dehydrating and decarbonating zones towards the flame and the 'clinkering zone' of the kiln. The resultant clinker is cooled immediately after leaving the kiln, either in planetary coolers fixed round the outside of the kiln shell, or rotary or grate coolers located immediately below the kiln.

A number of different types of cement-making processes are now in use; the rotary kiln is common to all but one of these methods, but the preparation of the kiln feed varies considerably. The main methods can be divided into four types known as the wet, the semi-wet, the semi-dry, and the dry processes. The various processes are summarized diagrammatically in Fig. 9.2. and described below.

In the wet process the raw materials are slurried in water, either in washmills for soft materials such as chalk, or in ball or tube mills for harder materials, and prepared and blended in such a way as to provide a kiln feed of the correct chemical composition and also to contain the minimum moisture consistent with pumpability. In the semi-wet process the slurry is partially dewatered by some form of filtration and the resulting filter cake formed into nodules which can then be fed to a pre-heater prior to entry into the kiln.

Raw material preparation in the case of the semi-dry and dry processes is carried out by crushing the various raw materials individually to about 25 mm and then feeding them, mixed to approximately kiln feed, to ball mills where they are homogenized and ground to a fine state. Final blending to the exact chemical requirements takes place in air-agitated silos. In the semi-dry process, the prepared raw meal is mixed with a small quantity of water in a rotating dish to form nodules which are then fed on to a moving grate. The hot exhaust gases from the rotary kiln pass through the grate and the bed of nodules, thereby drying and partly calcining them. In the dry process, the raw meal is fed either directly to a long rotary kiln or preheated in ducts by the hot exhaust gases from the kiln, the partly calcined material being separated from the gases by cyclones prior to entry into the kiln; this latter process is known as the suspension preheater.

The moisture content of the material fed to the kiln system clearly varies widely according to the process, from as much as 40 per cent or more in the wet process to nil in the dry process. The fuel consumption of the processes varies accordingly, being about 1400–1750 $kcal/kg$ of clinker in the case of a wet process to about 850 $kcal/kg$ of clinker in the case of a suspension preheater system. Although more economic from the aspect of fuel consumption, the semi-dry and dry systems can, if full advantage is to be taken of their high thermal efficiency, impose more rigorous limits on the acceptable levels of volatile constituents, since the preheating of the raw meal by the exhaust gases tends to allow dust and volatiles to

CEMENT MANUFACTURING PROCESSES

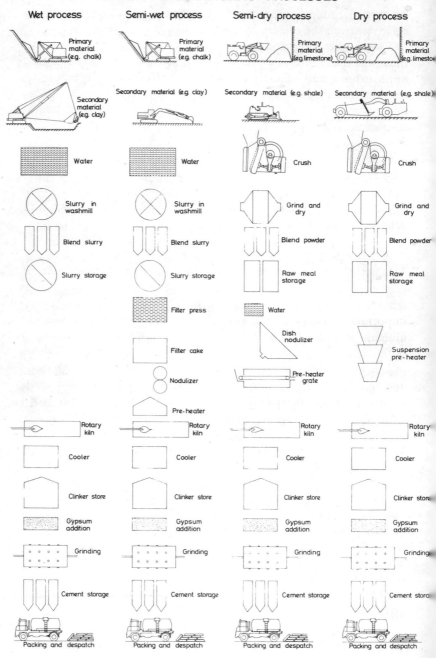

Fig. 9.2. Summary diagram of main cement manufacturing processes.

be carried back into the kiln by the raw meal rather than being removed from the system in gas-cleansing precipitators.

Two further processes may be mentioned here. In the first of these, the spray drier, the raw materials prepared in slurry form are atomized in a unit which is swept by the hot kiln exhaust gases; the resulting fine powder then passes into the kiln. The second process, using shaft kilns, is one of the very oldest methods and can be considered to be a development of the original limekilns. The kiln, consisting of a vertical cylinder lined with refractory bricks, is fed with an intimate mixture of raw materials and fuel, the formed clinker being removed from the base of the unit.

The layout of a typical modern cement works is illustrated on Fig. 9.3. This works, at Dunbar in Scotland, uses a semi-dry process to produce about a million tonnes of cement per year.

4. Chemical requirements of cement raw materials

The main prerequisite for manufacturing a satisfactory cement clinker is a raw meal of the correct chemical composition. In determining the correct composition three chemical ratios are of importance: the lime saturation factor, the silica ratio, and the alumina–iron ratio. The principal results of chemical analyses of both cement raw materials and cement clinker are quoted in terms of calcium oxide, silicon dioxide, aluminium oxide, and ferric oxide and it is to these values that the ratios relate. Under ideal conditions, the only loss which would occur from the system when combining the raw materials to form cement would be of any water present together with carbon dioxide from the carbonates; the various ratios in the clinker would therefore be the same as in the raw meal feed, since the relative proportions of the oxides remain the same. In practice, however, dust can be lost from the kiln and, if the firing fuel is coal, ash will be incorporated with the raw materials in the kiln so that the target ratios for the clinker are not generally the same as those for the raw meal; the difference between them will vary from kiln to kiln and from raw material mix to raw material mix. In the subsequent text the factors quoted will refer to the raw materials unless otherwise stated.

The lime saturation factor, commonly abbreviated to L.S.F., is given in B.S. 12 as

$$\frac{(CaO\%) - 0\cdot7(SO_3\%)}{2\cdot8(SiO_2\%) + 1\cdot2(Al_2O_3\%) + 0\cdot65(Fe_2O_3\%)}.$$

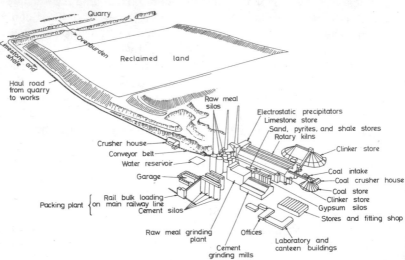

Fig. 9.3. Layout of typical modern cement works at Dunbar, Scotland.

The formula refers to cement and it is more normal when calculating the factor for raw materials not to consider the sulphate content. It is not unusual for the L.S.F. to be quoted in terms of percentages so that the equation becomes

$$\text{L.S.F.} = \frac{100(\text{CaO}\%)}{2\cdot8(\text{SiO}_2\%) + 1\cdot2(\text{Al}_2\text{O}_3\%) + 0\cdot65(\text{Fe}_2\text{O}_3\%)}$$

The silica ratio is defined as the ratio of the silica content of a material to the sum of its alumina and ferric oxide and is expressed as

$$\frac{\text{SiO}_2\%}{\text{Al}_2\text{O}_3\% + \text{Fe}_2\text{O}_3\%}, \text{ commonly written as } \frac{\text{S}}{\text{A}+\text{F}}.$$

The alumina–iron ratio is the ratio of the alumina content of a material to its ferric oxide content and is expressed as

$$\frac{\text{Al}_2\text{O}_3\%}{\text{Fe}_2\text{O}_3\%}, \text{ commonly written as } \frac{\text{A}}{\text{F}}.$$

The derivation of the L.S.F. formula is complex; it suffices to say that the aim in the manufacture of cement is to maximize the quantity of tricalcium silicate in the cement and that at a clinker L.S.F. of 1 the highest probability of realizing this aim is attained. The alumina–iron ratio determines the relative proportions of C_3A and C_4AF in the clinker; the silica ratio affects the proportions of silicates present. The alumina and iron oxide act as fluxes during the burning and hence affect the reaction, or burning, temperature but as this is also controlled by practical considerations, limitations are imposed on the ratios; cement plants in the U.K. operate at silica ratios in the range 1·5–4·0 and at alumina–iron ratios between 1·4 and 3·5. Under ideal conditions, the value of the clinker L.S.F. should approach 100 per cent (1·0 if the percentage notation is not used), but owing to practical considerations this figure will normally vary between about 90 and 100 per cent.

5. Physical requirements of cement raw materials

Although it is the chemistry of the raw materials which determine their overall suitability for cement clinker manufacture, the physical nature of these materials also plays a large part in process selection and the final design of a production plant. High inherent moisture contents of raw materials can give rise to handling problems; in dry process works, if the overall moisture is beyond the drying capability of the exhaust gases, auxiliary driers are required. On wet process plants, slurry viscosity can be a problem; although many raw materials produce a pumpable slurry at a moisture content of only 30 to 45 per cent, others require considerably more water to lower the viscosity to an acceptable level, thereby increasing fuel consumption at the kiln. Slurry moisture levels can be lowered by up to 5 per cent by means of additives, but these are often expensive.

Where a semi-dry process is used, it is essential that the kiln feed, after the addition of a little water, is capable of being formed into nodules which are not only strong enough to retain their shape while being fed to the preheater grate but can also withstand the thermal shock encountered when first meeting the hot kiln exhaust gases passing through the grate. The nodulization potential of raw materials appears to be related to the plasticity of the clay present, which itself seems to be a reflection of the bond strength between the layers of the clay minerals; weak bonding as in montmorillonite, vermiculite, or illite provides a better expansion and breakdown potential, leading to greater plasticity.

In the dry process, the temperature in the hottest part of a preheater can be as high as 1100 °C. Any raw materials containing significant amounts of minerals with a tendency to melt at these temperatures could be of restricted suitability for use in the process since they might encourage blockages in the preheater circuit.

6. Cement raw materials

Theoretically a Portland cement of good quality can be manufactured from any natural material or mixture of materials which, when blended together, would have an L.S.F. of about 90 to 100 per cent, a silica ratio of about 1·5 to 4·0, and an alumina ratio of about 1·4 to 3·5, and which, after grinding to a fine powder, could react to form a clinker at an economical burning temperature (about 1450 °C). In practice, the raw materials are normally a calcium-rich source, generally limestone, and a silica–alumina-rich source such as clay or shale. It is, of course, possible that a single rock-type could have a composition suitable for Portland cement manufacture. In practice it is extremely rare for a single rock-type not only to be chemically uniform but also to occur in sufficient quantity to support a cement works of any size. If such a material could be found it would constitute a one-component mix; normally mixes of at least two components are used. The major element in cement clinker is calcium and in a multi-component mix this will be supplied by the primary material; the vast majority of cement works therefore are sited at or near substantial deposits of limestone or chalk. Assuming that a calcium source has been located, a secondary material is needed and its required composition is determined by the L.S.F., $S/(A+F)$, and A/F of the primary material. Thus, if an argillaceous limestone has been selected as the primary material and its L.S.F. is somewhat in excess of 100 although the silica ratio and alumina–iron ratio are suitable, the secondary material would have to contain silica, alumina, and iron in roughly the same relative proportions as in the limestone so as to maintain acceptable silica and alumina iron ratios, but would need a low calcium content in order to lower the lime saturation factor; some form of clay or shale would therefore be sought. When combining two components, logic demands that only one of the ratios can be used as a target, the other two becoming thereby automatically fixed; should three components be used, two ratios can be fixed as targets; and all three ratios can be fixed as targets if four components are used. Should it be found that the addition of clay or shale, although adjusting the L.S.F., has altered the alumina–iron ratio to an undesirably high level, correction may be effected by incorporating an iron-rich material. For such a three-component mix, two of the ratios can be fixed; in this instance, the L.S.F. (the most important of the three), and the alumina–iron ratio would be chosen. Table 9.1 shows a typical calculation worked through with actual analyses and it can be seen that the final mix obtained would be quite satisfactory on basic chemical grounds.

The majority of cement works in the United Kingdom use only two-component mixes, and the bulk of the output is based on two main primary materials: Cretaceous Chalk in the south-east and Carboniferous Limestone in the north. The south-east of England has been a traditional location for the manufacture of cement for 150 years, much of the early development work being carried out on hydraulic limes and cements manufactured from chalk and Thames mud. Mud from the Thames is no longer used as a

TABLE 9.1

Typical calculation of raw mix based on materials available

Potential limestone primary material
Analysis: SiO_2, 8·00%; Al_2O_3, 1·23%; Fe_2O_3, 0·77%; CaO, 49·11%

This analysis gives a silica ratio of 4·0, an alumina–iron ratio of 1·6, and a lime saturation factor of 201. The essential required modification to this material is the reduction of the L.S.F.

Potential shale secondary material
Analysis: SiO_2, 59·78%; Al_2O_3, 18·41%; Fe_2O_3, 6·27%; CaO, 1·74%.

This analysis gives a silica ratio of 2·4, and a lime saturation factor of 1. Addition of the shale to the limestone will therefore produce a product with a silica ratio between 2·4 and 4·0, an alumina–iron ratio between 1·6 and 2·9, and an L.S.F. between 1 and 201; a potentially suitable kiln feed is therefore possible.

Using a target lime saturation factor of 96, which is a suitable value for the type of process being considered, the mix obtained is limestone, 87·7%; shale, 12·3%. This mix has an analysis which gives a silica ratio of 3·0, an alumina–iron ratio of 2·3, and a lime saturation factor of 96. The alumina–iron ratio could be reduced, if desired, by adding iron oxide fines.

Iron oxide fines
Analysis: SiO_2, 10·80%; Al_2O_3, 9·00%; Fe_2O_3, 72·40%; CaO, 0·00%.

Three components are available, limestone, shale, and iron oxide fines, so that two target parameters are required to obtain a mix; taking the target L.S.F. as 96 and the target alumina–iron ratio as 1·4, the mix obtained is: limestone, 87·1%; shale, 11·5%; iron oxide fines, 1·3%. The analysis gives a silica ratio of 2·5, an alumina–iron ratio of 1·4, and a lime saturation factor of 96. This mix would in theory be suitable as a basis for further investigations into the potential raw materials.

---

secondary material on account of its variability, having been replaced by either Gault Clay or Eocene clay. Despite the relatively high fuel consumption, and hence running costs, the wet process continues to be used in works based on the chalk because of the ease with which both this and the clay used can be broken down by agitation with water. This mode of preparation also permits pipeline transport and facilitates the separation, by screening, of flints which occur in the chalk; if allowed to remain, they could give rise to chemical variation and high grinding costs. The large works at Northfleet in Kent is situated on the south bank of the Thames and is therefore able to load clinker and cement for export directly into bulk ships as well as into rail and road transport for the home market; the primary and secondary raw materials are, however, situated 4 km and 8 km respectively from the works, the London Clay used as secondary material being on the north bank of the Thames in Essex. The clay, excavated by multi-bucket excavator, is

Fig. 9.4. Chalk winning in a modern cement works quarry (Northfleet, Kent). The mobile primary crushing plant with a throughput of 750 tonnes per hour is fed by a long-reach face shovel fitted with a 10-m³ bucket.

slurried and then pumped to the chalk quarries by means of a pipeline beneath the Thames. At the slurry preparation plant situated in the chalk quarry, the chalk, having been won by means of face shovels feeding large mobile crushers (Fig. 9.4), and is transported to the plant by conveyor where it is wet-milled and the flints screened out. It is slurried with the clay slip to form a kiln feed of the correct composition, which is then pumped by means of a further pipeline to the works.

The Carboniferous Limestone which is the mainstay of the various cement works in the north of England and Scotland is commonly mixed with Carboniferous shales for the manufacture of Portland cement. The wet process was originally used in these works, but more recent installations tend to be of the semi-dry and dry process types. The hardness of the Carboniferous Limestone makes it necessary to drill and blast the rock, and variations within the limestone itself often make it necessary to work more than one quarry bench at a time to produce a mix that will provide suitable grade stone and will also prolong the life of the reserves. After blasting, the limestone is normally loaded into dump trucks, either by face shovels or, where variation in the rock has made it necessary to blend a number of blasts, by mobile equipment such as front-end loaders, which then transport it to the primary crusher. At this stage, impurities, such as clay from cavities, can be screened out. Further crushers will then reduce the clean stone to a preferred maximum size of about 25 mm. According to its particular physical characteristics, the shale may be worked either by elevating scrapers, multi-bucket excavators, face shovels, or, if numerous hard bands occur, by drilling and blasting. Once again the material will be transported to crushers and broken down to a size of about 25 mm. The crushed limestone and shale are fed to stockpiles, whence they can be extracted together at predetermined rates and mixed and milled prior to blending in silos to the correct mix for use in the rotary kiln.

The varied geology of the British Isles is such that suitable cement-making materials are available in relative abundance. Furthermore, these materials tend to be in the form of a reasonably high-grade material such as fairly pure limestone and suitable low-grade forms such as shales low in calcium carbonate. Examples of the various geological horizons which are being used by the cement industry in the United Kingdom are given in Fig. 9.5.

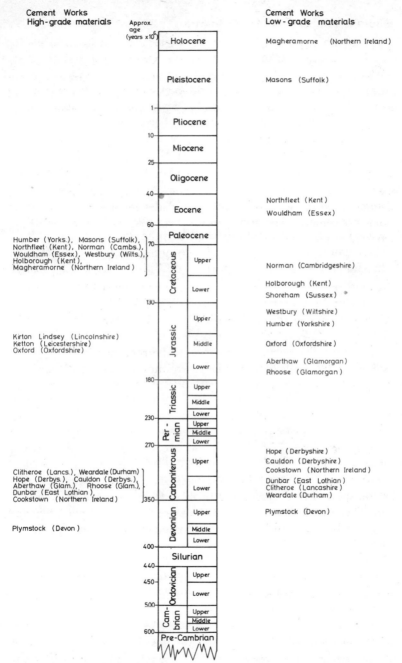

Fig. 9.5. Horizons exploited by selected United Kingdom cement works.

Geology and the cement industry

Elsewhere in the world, either the high-grade, or alternatively the low-grade, materials may be absent. For instance, at a works in the Orange Free State, South Africa, an area of extremely low-grade secondary limestone occurs in a region considered favourable for cement manufacture on the basis of market considerations. In this instance the high-grade material has been obtained by means of beneficiation of the low-grade limestone; methods employed not only entailed screening and froth flotation but also hand picking of unwanted materials. Even so, the variable nature of the deposit was such that a constant supply of beneficiated material could not be always guaranteed, so that it was necessary to maintain a store of high-grade limestone imported from a distant locality. At the other end of the scale is the Bahamas, again ideally situated not only for supplying the local market, but also Bermuda, the United States, the Caribbean, and Central America. The supply of coral limestone with a high calcium carbonate content in the Bahamas is virtually unlimited; secondary materials are, however, entirely absent and it is necessary to import them. Silica sand is brought in from Florida, bauxite from Jamaica, and iron concentrate is shipped from northern Canada; gypsum to be ground with the clinker to form the finished cement is brought in from Nova Scotia.

Unusual raw materials are exploited at a number of sites. At a works in Jalisco, Mexico, the volcanic deposits used as a secondary material are deficient in silica and to adjust the mix it is necessary to import diatomite from a location some distance from the works. At Puebla, also in Mexico, the moisture content of the materials is so high as to make it necessary to use a high proportion of basalt in the secondary material in order to facilitate handling the materials at the dry process works located there. The secondary materials at a works near Kuala Lumpur in Malaysia are shale and sandstone taken from the waste tips at a disused coal-mine some 15 km from the cement works. In Uganda, the lack of sedimentary limestone has made it necessary to use carbonatites as a source of primary material for cement manufacture; the high proportion of phosphate in the form of apatite does however mean that this igneous rock is not particularly suitable as a cement raw material.

In the most common combinations of raw materials, the primary material supplies the bulk of the calcium and the secondary materials the silica, alumina, and iron, but at a works in Perth, Western Australia, the primary raw material is coastal dune sand which contains not only all the calcium required for the kiln feed but also all the silica; the secondary material is therefore required only to supply alumina and iron oxide and this is achieved by using a ferruginous bauxite unsuitable for use in the near-by alumina industry.

7. Determination and proving of cement raw materials

Market surveys having indicated a favourable location for a cement works, it is for the geologist to determine whether or not suitable raw materials exist in the area. Under normal circumstances the site for a new works will be adjacent to its raw material source; for each tonne of cement produced about 1·6 tonnes of raw material plus its attendant moisture have to be transported to the works, and it is generally cheaper therefore to transport the finished cement rather than the raw material. The initial literature search and reconnaissance surveys are essentially similar to those carried out for any mineral or other raw material. Surface sampling for chemical analysis is undertaken during the reconnaissance stage of all materials which may be of potential use. Details of road, rail, and river access are noted, as are details of the water table and thickness of overburden. These last two points are of importance, since the potential costs of removing overburden and quarry pumping have to be considered in the cost equation of a new works project. Once potentially suitable areas have been located, a preliminary report is prepared indicating which materials would appear to be the most promising, and the likely quantities available, together with the difficulties that may be encountered during their extraction; a programme of detailed investigation would also be recommended. It may be possible at this stage to forecast that certain chemical constituents may be deficient in the potential raw mix, in which case the need for importing, say, iron oxide or silica sand would be indicated. Should the findings of the report be accepted as favourable, detailed investigations will then be undertaken; these normally include sampling by rotary core drilling.

Drilling operations have two aims: first to clarify the detailed geology of an area, and secondly to obtain enough samples for chemical analysis to enable detailed quarry development planning to be undertaken. Drilling patterns for interpretation of the geology are, by necessity, determined by the geology itself; for qualitative studies, however, it is preferred if a regular grid pattern can be used. In practice, financial considerations normally dictate that a compromise will be effected.

It should again be stressed that the quality of the output from the cement kiln is largely dependent on the quality of the raw material input since in the large majority of cases apart from crushing, grinding and blending, these raw materials are used in an essentially untreated form. Thus in order to maintain a continuous high-quality product, it is necessary to know in detail the chemical composition of the raw materials to be exploited and the variation within those raw materials. Good sampling, sub-sampling, and analytical techniques must be employed at all stages of the exploration. The highest possible core recoveries from rotary drilling are essential, as is accurate logging of the core. Analysis of borehole core should be carried out on the whole core rather than spot samples, unless a stage has been reached when it is known that a large degree of uniformity exists in the material, at which point it may be permissible to use equi-sized samples taken along a core at regular intervals. Cleanliness and accuracy in sample preparation for analysis are obviously of the utmost importance, and only those methods which can be shown to minimize sampling error should be used; coning and quartering of raw material samples is inadvisable, and the use of crushing or milling equipment which is not resistant to abrasion cannot to tolerated.

Once the geological field-work and drilling have been completed and all the required analyses performed, the geologist can describe the three-dimensional chemistry of the deposits under investigation. He should be able to recommend a course of action which will enable them to be worked in such a way as to maximize their life, while ensuring that at all times there will be materials available capable of being blended together to produce a raw mix of the required chemical analysis. The calculations leading to these recommendations are often complex and more than one best answer may apparently be available. Just as with base metal deposits, therefore, computer models of the deposits and simulations of the various possible extractive routes are now being used as an aid to optimizing reserve life and minimizing variation in the quality of the output. Some idea of the extent to which a deposit or deposits can be totally utilized can be obtained by calculating an overall analysis for each deposit, for instance the overall analysis for the limestone and the overall analysis for the shale deposit, and then combining these average analyses to give the required kiln feed mix; the proportions of each of the raw materials required to give a mix can then be compared with the total weights of the raw materials available.

The exact nature of the required mix is determined by cement technologists, chemists, physicists, and chemical engineers, both from the samples obtained for chemical analysis and also from samples obtained for physical and small-scale plant tests. For these last-mentioned tests, it is again the task of the geologist to ensure that the samples he provides are fully representative of the deposit.

Once a geologist has indicated the general lines along which a quarry should be developed, his plans will be worked out in practical detail by mining and quarrying engineers, bearing in mind that the equipment to be used may be required to blend as well as win and transport the raw materials.

A generalized scheme for raw materials investigations is given in Fig. 9.6, which also shows the interaction at various stages with other sections of the industry.

8. Harmful impurities in cement raw materials

A number of elements are detrimental to cement, and in certain cases to the manufacturing processes. These elements must be detected during the exploration stage. Of the impurities the most important, and certainly the most common, is magnesia. As much as 1·5 per cent of MgO can enter into solid solution in the several compounds present in cement clinker. Any remaining magnesia will remain uncombined in the form of the mineral periclase. This mineral is potentially dangerous since it hydrates slowly, eventually forming brucite, $Mg(OH)_2$. This change is accompanied by an expansion to twice the volume of the original periclase; the expansive forces generated by this change could crack a mortar or concrete. It is for this reason that a limit of 4 per cent magnesia in cement is laid down in B.S. 12; this limit is equivalent to an $MgCO_3$ content of about 5 per cent in the raw mix. One further chemical limitation laid down in the British Standard for Portland cement is for sulphur. Sulphur in the form of calcium sulphate (gypsum) is added to the cement during the final stages of manufacture, but sulphur also commonly occurs in the raw materials, often in the form of pyrite in the secondary materials; it can also be derived from the fuel. During the manufacturing process, sulphates will

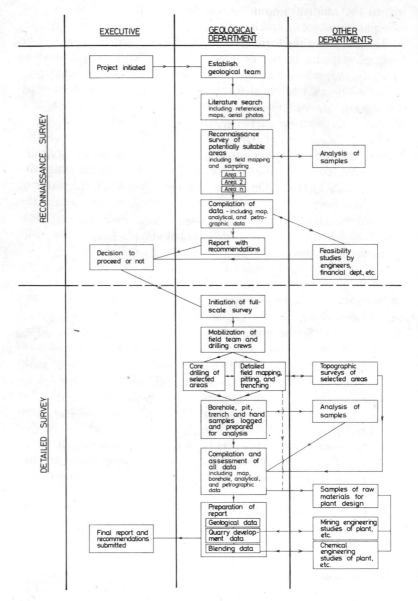

Fig. 9.6. Generalized scheme for raw materials investigation.

combine with alkalis present to form alkali sulphates and, according to the relative amounts of sulphate and alkalis present, the sulphur sometimes combines with calcium to form the sulphate, which appears in the form of the double salt $K_2SO_4 . 2CaSO_4$. The level of sulphate in the cement has to be strictly controlled to optimize strength and avoid unacceptable expansion of concrete.

The alkalis themselves also require consideration since they enter the silicates and aluminates as substitutes for calcium and there is therefore an increase in the effective L.S.F. and a modification of the C_3S/C_2S balance in favour of C_3S. Furthermore, pure K_2SO_4 formed in the clinker can react with added gypsum and water from the atmosphere to form the mineral syngenite $(K_2SO_4 . CaSO_4 . H_2O)$. Once this reaction has started, the water of crystallization present in the gypsum is sufficient to sustain the reaction; the effect is to give the cement a greater tendency towards 'air-setting' if stored under conditions of high humidity.

Other elements which should be noted include phosphorus, the effect of which on the stability of the various forms of calcium orthosilicate has been noted above. Although maximum levels have to be controlled to avoid loss of hydraulicity in the cement, 0.2 to 0.3 per cent P_2O_5 is beneficial as it helps to resist the change of belite to γ—C_2S, the non-hydraulic low-temperature form. Manganese acts in a manner analogous to iron and when significant quantities are present should be added to the ferric oxide when calculating L.S.F., $S/(A + F)$, and A/F. Fluorine can occur in raw materials usually in the form of fluorspar, CaF_2. It is also sometimes added in small quantities to raw material mixes as a mineralizer to facilitate clinkering at a lower temperature than would otherwise be required. Strict control of the fluorine level in clinker has, however, to be exercised to ensure that the cement characteristics are not altered. If it is allowed to rise to higher levels, lower strengths can result.

It is clear that the levels of the minor elements have to be carefully assessed and they may, in the limit, restrict the viability of any particular deposit. Not only does their interaction mean that the combinations of these minor elements have to be considered but since a number of them are volatile at sintering temperatures they may also dictate the type of cement manufacturing process chosen to exploit the deposit. Each case has to be treated on its own merits and the advice of both plant and cement technologists has to be sought in the final assessment and evaluation of potential raw materials located by the geologist.

9. Blending of raw materials

The duties of a cement geologist overlap with those of a cement chemist in determining the chemical suitability of raw materials; they also overlap with those of a chemical engineer when considering the blending of the raw materials to form a uniform kiln feed. Apart from crushing and grinding, cement raw materials are normally used in their as-dug state and the deposit from which they are won can be considered as a large stockpile.

By recommending the manner in which the deposit is to be worked, the geologist has already involved himself in the blending function of the process. This involvement is even more apparent when it is remembered that, in general, one of the main functions of a recommended quarry development plan is to ensure that at all times materials with grades both above and below kiln feed requirements are available for blending together.

As can be seen from the summary of blending methods given as Fig. 9.7, methods of raw materials blending can be divided into two types: first, those using specially designed blending equipment; secondly, those using the equipment actually used for winning the materials. The first type of blending method is properly the responsibility of the chemical engineer, although he will work closely with the geologist when designing the plant, since it is the geologist who will be able to supply the chemical and physical variances of the raw materials entering the system. It will therefore be the geologist who will, for example, advise the engineer that the high variance in the silica of a certain deposit is more complicated than the chemical analysis would tend to suggest, since he has determined that it is not only due to large chert fragments, but also to very small secondary crystals of quartz, and also the presence of clay minerals; the chemical engineer would then have to design a system to homogenize three very different mineral types if uniform silica distribution is to be obtained and segregation avoided.

The construction and running costs of all equipment designed for the specific purpose of raw material blending are very high and if better blending can be achieved by modification of some activity already being carried out to obtain the raw materials, a substantial saving may well ensue. The concept of the raw material deposit as a

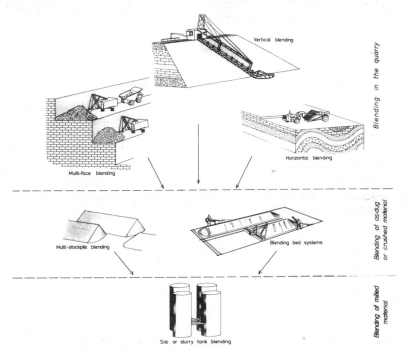

Fig. 9.7. Summary of raw material blending methods.

large stockpile suggests ways in which this can be done. The use of layered stockpiles has been fairly common in the extractive industries for some years, the smoothing of a variable input being achieved by building a stockpile from a series of layers of material and then extracting by taking successive slices through the stockpile so that each slice contains all the layers of material. This concept is directly applicable to the winning of stone from a geological 'stockpile' of raw material *in situ*. Taking a series of relatively similar slices through a deposit can be achieved in a number of ways. Long single faces can, for example, be blasted and the loading shovels progressively moved backwards and forwards along the pile of fallen stone so as to achieve a mix; for soft rock, draglines or multi-bucket excavators can be used in a similar manner. Such methods can be expensive if large shovels are required and are impracticable if electrically driven equipment is in use. Even if very mobile equipment such as the front-end loader is used there may still be problems, especially in the case of blasted materials in wet climates, where large quantities of stone lying on a quarry floor for long periods may absorb potentially unwanted moisture. A very successful method of taking repeated slices through a deposit is by means of an elevating scraper. This equipment is capable of skimming off a thin layer from the surface of the deposit, thereby covering quite a large area before obtaining a full load; it also has the advantage of being able to transport the material to the raw material preparation plant and, by discharging its load slowly while on the move, of building up crude layered stockpiles. With a deposit which varies horizontally, scrapers would be operated by taking thin horizontal layers off the top of the deposit and slowly lowering this top surface; with vertically variable strata, scraping off an incline which cuts all the strata would be used. Equipment of this type has been successfully used on materials ranging from glacial till to Cretaceous chalk and Carboniferous shale.

In the past the tendency for works and quarry management to pick the 'eyes' out of a deposit has been as prevalent in the cement industry as in other extractive industries. Although the selective quarrying of the more easily used portions of a deposit is now not normally practised, the technique, if properly controlled, can be put to good advantage. The first requirement is for the geologist to be able to predict with a high degree of accuracy the chemical and physical nature of all parts of the deposit. If the variation in a quarry is due to the presence of a large number of fairly distinct but separately uniform beds of rock, then the prediction problem is largely capable of solution by detailed geological mapping. Where, however, the variation is within, say, a massive limestone, conventional geological methods frequently cannot be applied and the geologist has to resort to mathematical and statistical methods. Such methods as trend surface analysis and moving averages have been used within the industry, as have traditional statistical methods and, more recently, geostatistics. All these techniques as applied to cement raw materials tend to have drawbacks, although geostatistics is showing more promise than most methods. The chief problem is the scale upon which detail is required if the techniques are to be of use; predictions of the chemical analysis or physical properties of blocks of stone as large as 200 tonnes to an accuracy of possibly ± 2 per cent of the true value are required. Many deposits are not, however, too complicated, and acceptable mathematical models can be constructed. Once the analyses of the various blast-blocks have been estimated, panels can be selectively worked with the aim of minimizing the variation in the quarry product while maintaining a composition as near the required kiln feed as possible. Multi-panel working will invariably involve the use of a fleet of dump trucks. In order to avoid delays at the crusher it is necessary to have flexible vehicle operation. Rather than having a fixed number of dump trucks operating with each shovel, the trucks are directed to the panel from which stone is required at that time in order to maintain the mix proportions. This procedure can be handled manually in fairly simple operations, but in large quarries it may be necessary to install a mini-computer and automatic signalling equipment to maintain the correct flow of stone from different parts of the quarry.

When recommending development of a deposit the geologist must therefore work closely with specialists in quarrying practice and chemical engineering. The best choice of a quarry development scheme and extractive equipment can well result in considerable savings in both capital expenditure and running costs.

10. Environmental problems

In common with other extractive industries, the cement industry is faced with the problem that its basic raw materials are largely located in areas of outstanding natural beauty. In Great Britain, for instance, a large part of the Carboniferous Limestone is in National Parks. The chalk areas in the United Kingdom, on the other hand, occur in the most densely populated part of the country, the south-east. The environmental problems associated with winning of the raw materials, dust, and the noise of blasting and heavy vehicles, are common to all the extractive industries and every effort is made to counter them. Financial considerations tend, wherever practicable, to lead to extraction in areas where overburden is thin and workings can be as deep as possible. From one aspect this means that large spreading quarries can be avoided, but it also means that when they are worked out deep scars are left. Infilling of old quarries with household refuse has become commonplace, although when the water-table has been breached great care must be taken if the local aquifer is not to be polluted. In some areas, market considerations have caused cement works to be so sited as to exploit deposits with high overburden ratios. In such cases progressive reclamation procedures can be used to return the land to its original use, as at the Dunbar Cement Works east of Edinburgh (Fig. 9.3). Where little or no backfilling can take place, abandoned cement works quarries have been used as recreational areas. Cement works tend to be fairly close to areas of high population density, and are therefore ideally suited for such purposes. What can be achieved can be seen at the famous Butchart Gardens on Vancouver Island, British Columbia, where an abandoned cement works quarry over 12 ha in area has been transformed into one of the most spectacular botanical gardens in the world. Although quarry development is largely dictated by geological, chemical, and physical considerations, environmental impact is a factor of increasing importance.

11. Conclusion

The cement industry depends almost entirely upon natural raw materials, and the geologist has a particularly important role to play in ensuring that suitable supplies are available in the long term. The nature of the industry and the manufacturing process is such that the geologist must work closely with both chemists and engineers in supplying practical advice. The geologist in the cement industry is also called upon to give professional advice on many other matters, apart from raw materials. Water boreholes for works supplies, foundation problems for heavy fixed plant, data for reports made in compliance with Government regulations, and many other varied tasks all fall within his scope. As the demand for cement increases, so does the desire for production units close to new markets; potential raw materials in these areas are often of marginal quality. The geologist's task is therefore becoming more exacting and he is continually seeking to improve his techniques for finding, proving, and developing these raw materials and, together with his colleagues from other disciplines, to improve the efficiency of one of the world's most basic industries.

Acknowledgements

The author wishes to thank the Chairman and Directors of the Associated Portland Cement Manufacturers Ltd for permission to publish this contribution and to reproduce the photographs incorporated in Figs. 9.3 and 9.4, and to acknowledge the constructive comments and assistance of numerous colleagues in the preparation of the text.

References

AMERICAN SOCIETY FOR TESTING MATERIALS (1973). *Annual book of A.S.T.M. standards, Part 9. Cement; lime; gypsum (including manual of cement testing).*

BOWEN, N. L., and GREIG, J. N. (1924). The system: Al_2O_3, SiO_2. *J. Am. Ceram. Soc.* **7**, 238–54.

BRITISH STANDARDS INSTITUTION (1958). B.S. 146: 1958. *Portland blast-furnace cement.*

—— (1958). B.S. 1370: 1958. *Low heat Portland cement.*

—— (1968). B.S. 4246; *Low heat Portland blast-furnace cement.*

—— (1970). B.S. 4627: 1970. *Glossary of terms relating to types of cements, their properties and components.*

—— (1971). B.S. 12: Part 2: *Portland cement (ordinary and rapid hardening) Part 2, metric units.*

—— (1972*a*). B.S. 890: 1972. *Building limes.*

—— (1972*b*). B.S. 915: Part 2: 1972. *High alumina cement. Part 2, metric units.*

—— (1972*c*). B.S. 4027: Part 2: 1972. *Sulphate-resisting Portland cement. Part 2, metric units.*

CEMBUREAU (1972). *World cement directory.* The European Cement Association, Paris.

LEA, F. M. (1970). *The chemistry of cement and concrete.* Edward Arnold, London.

RANKIN, G. A., and WRIGHT, F. E. (1915). The Ternary system, $CaO–Al_2O_3–SiO_2$. *Am. J. Sci.* **39**, 1–79.

10. Groundwater

1. Introduction

Water is a basic necessity of life and, as such, it is probably the world's most precious resource. While this is generally appreciated, it is less widely realized that groundwater represents a very large proportion of the world's readily available fresh water reserves and is therefore one of its most valuable assets. Only about 3 per cent of the total volume of water in the hydrosphere is fresh water but some 75 per cent of this occurs in the polar ice-caps. Of the remainder, most is stored as groundwater, groundwater storage being greater than that in lakes and rivers by a factor of at least thirty and possibly considerably more (Anon 1971). The magnitude of the difference emphasizes the potential of permeable rocks as sources of water.

Groundwater has the very considerable advantages that it is cheap to develop, it has a uniform quality and temperature, it is less susceptible to pollution and, because of the large storage capacity of permeable rocks, it has a more uniform distribution in space and time and hence supplies are less likely to be affected during droughts.

In the past, groundwater has supported irrigation in many agricultural economies, leading to increased wealth and, in some cases, to the creation of more diversified urban and industrial communities. There are large amounts of uncommitted groundwater storage in many parts of the world at present underdeveloped, and these reserves afford considerable potential for economic growth and improved standards of living and hygiene. Much of the world's groundwater resources occur in Pleistocene and Recent alluvial deposits in lowland regions and are favourably placed for economic development.

At a variable, but relatively shallow, depth below the earth's surface, the rocks are saturated with water. The upper boundary of this saturated zone is referred to as the water table and it is the water in this zone that is called *groundwater* (Fig. 10.1). Above the water table there is an unsaturated zone (or zone of aeration) in which the rock interstices are filled with varying proportions of water, water vapour, and air.

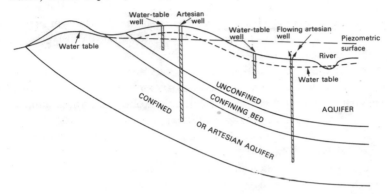

Fig. 10.1. Unconfined and confined aquifers.

Groundwater is mainly replenished by rainfall and by leakage from rivers and lakes. The water derived from these sources, referred to as *meteoric water*, infiltrates through the unsaturated zone to the water table thereby replenishing groundwater storage. Groundwater forms part of the hydrological cycle, slowly moving through the ground in the direction of the hydraulic gradient before discharging from springs or seepages into rivers or lakes or directly into the sea. Part of the water stored in rocks is the remnants of water trapped in sediments at the time of their deposition and is referred to as *connate water*. It is generally saline and occurs at depth, having been flushed out of rocks near the surface by circulating meteoric water. A further source is *juvenile water*, which is of magmatic or volcanic origin.

Groundwater plays an important role in geological processes. It is involved in rock weathering, in the leaching and cementation of rocks, and in the redistribution of elements and hydrocarbons leading to their concentration as ore deposits and oilfields. Knowledge of the occurrence of groundwater and the factors controlling its movement are important in civil engineering works: its movement affects the stability of excavations and tunnels;

extensive drainage provisions are often necessary before minerals can be worked in mines and quarries or before large engineering works can be undertaken, and its rate of movement must be assessed in designing reservoirs or other hydraulic structures. Geothermal energy may be transported by subsurface fluids; saline connate waters may be regarded as low-grade ore deposits, and some have been exploited. Thermal and saline waters are used for therapeutic purposes, although less so than in the past. Industrial wastes are disposed of in rocks at depth, additional storage being created by injection under pressure.

Some of these various roles are referred to in other chapters but the main purpose of this chapter is the consideration of ground-water as a mineral resource for use as a source of water supply; this is the principal task of the hydrogeologist.

2. The nature of aquifers

An aquifer is a rock which is sufficiently permeable for water to move through it at rates allowing economic development. The introduction of cost into the definition means that a rock which is an aquifer to a farmer may not necessarily be so to an industrialist, but such a definition emphasizes that an aquifer contains a mineral—groundwater—and that economics enters into its development. An aquifer fulfils three commercially valuable functions: it stores water, it transmits water, and, if it is an unfissured rock, it will filter and purify water. Advantage is taken of these three functions wherever groundwater is used but, in recent years as aquifers have been more intensively developed, detailed consideration of these functions has entered into the management of groundwater resources.

There are two types of aquifer: unconfined aquifers in which the upper surface of the saturated zone, that is the water table, is open to the atmosphere and is therefore at atmospheric pressure; and confined, or artesian, aquifers, in which groundwater is confined by overlying relatively impermeable deposits at fluid pressures greater than atmospheric pressure. When wells penetrate a confined aquifer the water rises above the base of the confining bed to a level determined by the hydrostatic pressure. The imaginary surface defining this is referred to as the piezometric or potentiometric surface. In some situations water from an artesian aquifer flows out at the surface (Fig. 10.1).

Fig. 10.2. Cores of cemented shell sand from the Bahamas showing high primary porosity.

The porosity of a rock is a measure of its void space and is expressed as a percentage of the total volume of the rock. Thus, if a rock is saturated, porosity indicates the total water content. Primary porosity represents the rock interstices created when the rock was formed (Fig. 10.2), whereas secondary porosity depends upon processes which have modified the rock subsequent to its formation. The ability of many rocks to transmit water, particularly limestones, depends upon the development of secondary porosity in the form of fissures, fractures, and joints. With increasing depth the frequency and size of such openings decrease. Although groundwater may be encountered in mines at depths of more than 1000 m and will flow through sedimentary rocks at much greater depths, the amounts involved and the rates of flow are normally small.

Groundwater is usually abstracted from the upper 100 m of the saturated zone. Many wells of considerably greater depth than this are in use for water supply, particularly wells drilled into artesian aquifers although, of course, water is not pumped from the depth of the aquifer itself as advantage is taken of the artesian pressure; from an economic point of view 100–200 m would be a generally acceptable maximum pumping lift.

The important factor in groundwater development is not the total water content of a rock but the amount of water that can be drained by gravity, that is the water-yielding capacity, which is referred to as the *specific yield*. The proportion retained in a rock by molecular attraction and capillarity is the *specific retention*; the sum of the specific yield and specific retention is equal to the porosity. Reference to groundwater storage is invariably related to specific yield and not total water content.

TABLE 10.1

Representative values for porosity, specific yield, and hydraulic conductivity

	Porosity (%)	Specific yield (%)	Hydraulic conductivity (m/day)
Gravel	20–35	15–30	10–>1000
Sand	35–40	15–30	1–100
Sandstone	5–25	⩽15	10^{-3}–100
Limestone	1–10	<5	10^{-4}–100
Chalk	20–50	1–5	10^{-3}–100
Clay	>45	Negligible	10^{-7}–10^{-3}

The relationship that specific yield bears to porosity depends upon the grain size and the shape, distribution, and continuity of the pore spaces (Table 10.1). In a coarse gravel the specific yield may approach the porosity with a value ranging up to 25 or 30 per cent, whereas in a fine-grained deposit, such as chalk, the typical porosity is some 30 to 40 per cent but the specific yield is as low as 1 or 2 per cent.

The *storage coefficient* of an aquifer is defined as the volume of water released from (or taken into) storage per unit surface area of the aquifer per unit change in hydrostatic pressure. The term should be applied only to confined aquifers; the equivalent term for unconfined aquifers is the *specific yield*. As already mentioned, this indicates the storage available to wells provided that gravity drainage is complete (Fig. 10.3a). But when a confined aquifer is pumped it may remain fully saturated. As the hydrostatic pressure declines, the aquifer is compressed, reducing the porosity slightly, and simultaneously the water expands to a limited extent. By these processes water is released from storage, the amount being related to the storage coefficient. Typical values are about 10^{-4}, indicating that considerable changes in hydrostatic pressure are required over extensive areas before large amounts of water are obtained from a confined aquifer. The specific yield for a gravel deposit may be as much as 500 times greater than the coefficient of storage for a typical confined aquifer.

The *permeability* of a rock measures the ease with which water moves through it. The velocity of the flow of water through a natural porous medium is proportional to the hydraulic gradient and a coefficient referred to as the hydraulic conductivity. This statement, known as Darcy's Law, may be simply expressed as

$$Q = \frac{K(h_1 - h_2)A}{L} \text{ or } KiA,$$

where Q is the flow rate, K is the hydraulic conductivity, $(h_1 - h_2)$ is the head loss over distance L, i.e. the hydraulic gradient i, and A is the cross-sectional area of the porous medium through which flow occurs. The hydraulic conductivity is defined as the rate of flow of groundwater through a unit cross-sectional area under unit hydraulic gradient (Fig. 10.3b). Values for good aquifers are greater than 30 metres/day (1 m/day$\approx$1 darcy) and for many gravels exceed 500 m/d, whereas 10^{-5} m/d is more typical for unweathered clay (Table 10.1). The transmissivity of an aquifer is the product of the hydraulic conductivity and the saturated thickness and relates to the ability of the aquifer to transmit water through its entire thickness.

The hydraulic conductivity is a function of both the nature of the medium and the fluid properties, viscosity and specific weight. The permeability of the medium itself is related to pore diameter and such depositional factors as porosity, shape, and packing of the grains and their size distribution. It is referred to as the *intrinsic permeability* and has dimensions of area, not velocity. Values are extremely small when expressed as cm^2 and therefore, the darcy (which equals $0.987 \times 10^{-8}cm^2$) is the practical unit adopted, although square micrometre ($10^{-12}m^2$) is also used.

An important point is that the velocity (Q/A), derived from the above equation expressing Darcy's Law, is an apparent velocity for it is the flow through the total cross-sectional area A, whereas in actual fact flow takes place only through the pores of the rock. To determine the average velocity, porosity (n) has to be introduced

$$V = \frac{Q}{nA} \text{ or } \frac{Ki}{n}.$$

An aquifer is said to be isotropic with regard to permeability if the permeability is the same in all directions and anisotropic if values differ in different directions. If the permeability is constant throughout the aquifer it is said to be homogeneous. In reality most aquifers are anisotropic and non-homogeneous but on the large scale homogeneity and isotropy are commonly assumed in order to facilitate the solution of groundwater flow problems.

Typical velocities of groundwater flow range from 1 or 2 m/day to 1 or 2 m/year, although in very fissured limestones maximum rates are similar to those in rivers and turbulent flow conditions obtain rather than laminar flow which is more usual in porous media.

The world's aquifers are mainly unconsolidated alluvial sands and gravels in deltas, in river flood-plains, or infilling buried valleys. Sandstones and limestones are also important and volcanic rocks may be very permeable where vesicles, joints, or cracks have developed. Igneous and metamorphic rocks are generally poor aquifers although small amounts of water can be obtained from the upper weathered zone, which may attain thicknesses of the order of 50 m.

The groundwater level in an aquifer indicates the volume of water in storage. Fluctuations of the level reflect changes in storage related to recharge to, and discharge from, the aquifer. A cyclic pattern of levels is usual in many regions, for example in north-west Europe. Natural recharge occurs in the winter causing levels to rise, but during the summer evaporation is high, often exceeding rainfall, and infiltration is small or negligible with the result that levels steadily decline as the winter storage drains to natural outlets, minimum levels being attained in the autumn. In confined aquifers,

---

Fig. 10.3. Definition of aquifer properties: (a) Storage coefficient of confined and unconfined aquifers; (b) Hydraulic conductivity and transmissivity. Hydraulic conductivity K is the quantity of water flowing in unit time through unit cross-sectional area under unit hydraulic gradient ($m^3 d^{-1} m^{-2}$ or m/d). Transmissivity T is the quantity of water flowing in unit time through unit width of a saturated aquifer under unit hydraulic gradient. It is equal to K m ($m^3 d^{-1} m^{-1}$ of m^2/d). (Fig. 10.3a is 'after Todd' (1959) by courtesy of John Wiley and Sons Inc.)

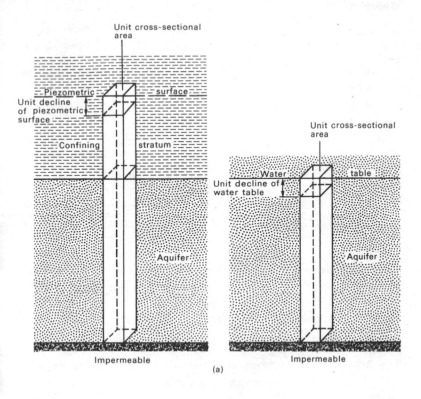

(a)

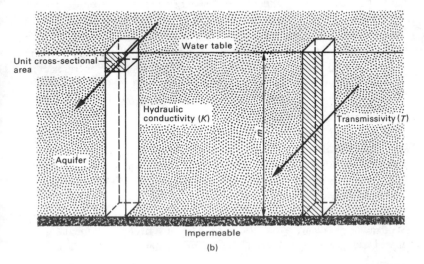

(b)

231

water levels respond to changes in atmospheric pressure, earthquakes, the tidal cycle in coastal areas, and artificial loading such as that caused by the passage of trains. If a groundwater resource is being over-developed water levels will steadily fall.

Two forms of groundwater storage may be recognized: replenishment and permanent storage. The water that replenishes an aquifer seasonally, or less frequently in many arid regions, is referred to as replenishment or temporary storage. It occurs in the zone of active water circulation and most of it is discharged to rivers or the sea before the next recharge season, although the large storage capacity of aquifers tends to balance variations of supply from year to year. Permanent storage is the water lying below the principal outlets from an aquifer. The volume in major aquifers is usually considerably larger than the average seasonal input.

In contrast to temporary seasonal storage, which flows relatively quickly in the zone of active circulation, most of the water representing permanent storage is moving very slowly in deeper regional flow-systems. Under natural conditions this water may be flowing to outlets by upward leakage through relatively impermeable confining beds.

3. Assessment of groundwater resources and development

Assessing the groundwater resources of an area entails identifying the main aquifers and the source and amount of natural recharge, as well as estimating the volume of storage that can be developed. Variations in thickness and lateral extent of an aquifer and its associated less permeable beds have to be determined and the aquifer boundaries defined. 'Boundaries' are features that influence groundwater flow, such as faults (which may either restrict or encourage flow), facies variations (which may reflect changes in permeability or storage), or rivers (which may act as recharge sources or may have essentially impermeable beds).

The extent of an aquifer is determined by geological mapping, examining existing well records, drilling exploratory boreholes, and making geophysical and aerial surveys. The nature of the aquifer and its hydraulic properties are assessed from field examination and the collection of rock and borehole core samples for the measurement of permeability and specific yield in the laboratory, although, as discussed below, hydraulic properties are preferably measured by test pumping wells. If possible, groundwater-level maps are made

Fig. 10.4. Spring discharging at contact between Lincolnshire Limestone (above) and Upper Lias Clay (below).

and the fluctuation of the water levels determined. The outflow from the aquifer is measured (Fig. 10.4), water samples are chemically analysed and potential sources of pollution identified.

The form such a study takes depends very much upon the nature of the problem, the time available, the availability of existing data and, of course, economic considerations. An initial reconnaisance survey is commonly followed by exploratory drilling and test pumping and finally the development of any resources identified.

Assessment of the natural recharge to an aquifer often requires the solution of a general water balance equation of the form

$$P = E + S_R + G_R + \Delta S + U,$$

where P denotes precipitation, E evaporation (including transpiration by vegetation), S_R the direct surface run-off and interflow (that is water which flows to rivers through the unsaturated zone, mainly in the soil and sub-soil, without reaching the saturated zone), G_R the groundwater discharge, or base flow, ΔS the change in storage, and U the subsurface flow into or out of the catchment.

If the balance is considered over an extended period of time, say

233

Groundwater

10 years or more, the change in storage becomes insignificant. Where permeable rocks crop out and there is no surface run-off, natural recharge is the difference between precipitation and evaporation. If river flow data are available, analysis of the form of the river hydrograph can identify the three basic components, direct surface run-off, interflow, and groundwater discharge (Fig. 10.5). If the geology of the catchment is relatively simple, and for example includes only one aquifer, then the groundwater discharge can be related to the infiltration into the permeable deposits.

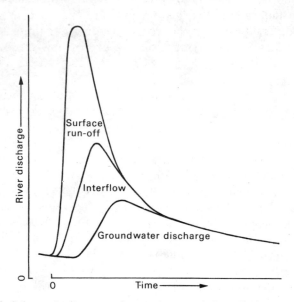

Fig. 10.5. Schematic diagram of components of river discharge.

In arid and semi-arid regions groundwater resources may be mainly replenished by influent seepage from intermittent streams, perhaps fed by storm run-off from mountainous regions. In these circumstances the average groundwater resources may be estimated from variations in groundwater levels or the flow of water through the aquifer can be assessed from Darcy's Law expressed in the simple form

$$Q = TiW,$$

where Q denotes the rate of flow, T the transmissivity, i the hydraulic gradient, and W the width of aquifer through which flow takes place. The resources held in permanent storage are estimated by multiplying the volume of the saturated aquifer by the storage coefficient or specific yield.

The resources which are naturally available may be supplemented artificially. The process, referred to as artificial recharge, involves the recharge of an aquifer with surplus river water through infiltration basins or wells, or by spreading the water over permeable ground or by modifying stream beds and adjacent areas to encourage infiltration. Various methods are employed as appropriate in different parts of the world including the United States, Europe, and Israel. During the recharge process low-grade waters are improved in quality and in some schemes this is the prime purpose.

Geophysical methods are used increasingly in the exploration for groundwater and its subsequent development. During the exploration stage surface techniques, such as electrical resistivity, seismic, gravity, and magnetic methods, are applied to identify geological conditions favourable for the occurrence of groundwater. Resistivity and seismic methods may be employed, for example, to determine the extent and thickness of gravel deposits or locate a buried glacial channel containing extensive thicknesses of sands or gravels. The depth to a saline interface can be mapped by resistivity methods.

After a well (or borehole) has been drilled, borehole logging is used to determine the nature and properties of the rocks exposed in the well and the water they contain and also of the water column in the well (Fig. 10.6). Electrical resistivity logs reveal the nature of the strata and are valuable for correlation. These logs can be supplemented by natural gamma-ray logs where clays or mudstones are included in a sequence, advantage being taken of the relatively high radioactivity of argillaceous rocks. An electrical resistivity log of the well fluid reflects the chemical quality of the water. Levels of water inflow to a well may be indicated by temperature logs or by a sensitive flow-meter. Contributions made by different aquifers to the yield of a well can be assessed by a flow-meter in conjunction with a caliper log, which records changes in the diameter of a well (Fig. 10.6b). Maximum benefits are obtained from borehole logging when the various logs are interrelated and studied with all the available well data. Interpretations can be assisted by direct inspection with an underwater television camera.

Nowadays groundwater is usually developed by means of wells, although in the past horizontal tunnels were sometimes driven to

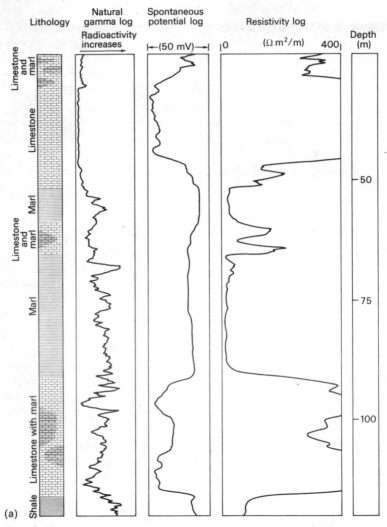

Fig. 10.6. Applications of borehole logging: (a) Comparison of gamma ray, spontaneous potential, and electrical resistivity logs with a lithological log; (b) the identification of a water-bearing fissure by borehole logging.

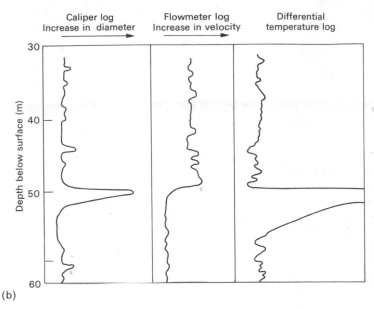

(b)

supplement the yield from individual wells. For example, London derives part of its water supply from a system of 48 pumping stations (each including more than one well) and 18 km of tunnels (or adits) in the Chalk. For many centuries underground tunnels (referred to as qanats) have been used in Iran, Afghanistan, and other parts of the Middle East; they were driven to intersect the water table and act as collecting galleries. The average length is a few kilometres but some are many kilometres long. Volcanic rocks in the Canary Islands are exploited by large-diameter wells interconnected with galleries, many of which are over a kilometre long. Where saline intrusion is a problem and excessive water-level drawdown is undesirable, there is considerable advantage in deriving water from tunnels driven near the top of the freshwater lens (Figs. 10.12). One such tunnel in Hawaii, about 500 m long, yields over 200 Ml/d from very permeable basalts (Peterson 1972).

A major well capable of yielding 50 litres or more of water per second will have a diameter of about 450 or 600 mm. Wells for small agricultural supplies, amounting to 2 or 3 l/s, may have diameters of the order of 150 mm. If the aquifer is unconsolidated a well screen will be necessary to prevent its collapse into the well. The screen is a casing with perforations or slots which allow water

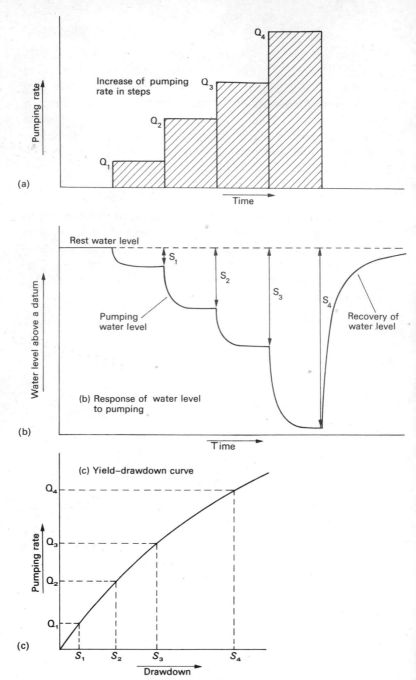

Fig. 10.7. The step-drawdown pumping test: (a) increase of pumping rate in steps; (b) response of water level to pumping; (c) yield–drawdown curve.

to flow into the well but prevent coarser particles in the aquifer from entering with the water. A well is 'developed' by pumping which removes the finer particles from the aquifer in the vicinity of the well, thereby increasing the porosity, permeability, and effective diameter of the well. When the aquifer is a fine-grained uniform sand, a gravel pack is placed between the screen and the well to stabilize the aquifer and prevent sand entering the well. The slot size of the screen and the grain size of the pack are carefully chosen in relation to the aquifer to ensure long-term optimum yields from the well.

After a well has been drilled it should be pumped to determine its yield and the properties of the aquifer. The yield test involves pumping at three or more rates until the pumping water level becomes relatively steady. This is referred to as a step-drawdown test (Fig. 10.7). The relationship between the pumping rate and the decline of water level in the well at each rate defines the yield–drawdown curve (Fig. 10.7), and from such a curve the permanent pumping plant can be designed and an estimate made of the pumping costs.

The aquifer properties may be determined if one or more observation boreholes are available in the vicinity of the pumped well. The well is usually pumped at a constant rate and the decline of the water level in the observation boreholes is measured as the cone of depression develops (Fig. 10.8). The aquifer properties, i.e. trans-

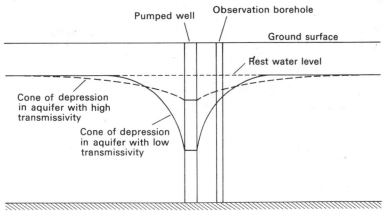

Fig. 10.8. The different shape and size of a cone of depression in an aquifer with a low and a high transmissivity. Pumping rate and length of pumping period, etc. are the same in both cases.

Groundwater

missivity and storage coefficient, are calculated from standard formulae. The existence of hydrogeological boundaries can be inferred by studying changes in the rate of decline of water levels (Fig. 10.9). When the aquifer properties are known, the consequence on groundwater levels of pumping at various rates for any length of time can be estimated. Aquifer properties calculated from field pumping tests are representative of relatively large volumes of the aquifer and are to be preferred to laboratory measurements using small samples of rock which are liable to sampling error.

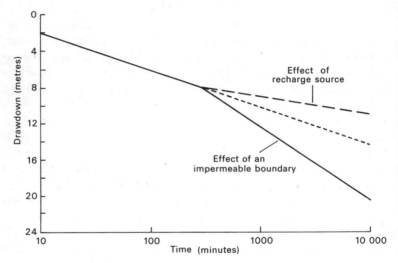

Fig. 10.9. The rate of change of drawdown in an observation borehole due to pumping a near-by well, and the effect of an impermeable boundary and a recharge source.

4. Consequences of groundwater development

During the early stages in the development of the groundwater resources of an aquifer, the volume of water pumped is small in relation to the annual natural recharge and the total storage. But, as abstraction increases to become a significant proportion of the available resources, the consequences become apparent through declining water levels, reduction of river flows, saline intrusion in coastal areas, and, in some situations, subsidence of the ground surface. Because of the large volume of water in storage and its slow rate of flow, the effects of abstraction are not felt immediately throughout an aquifer. One of the important tasks of a hydro-geologist is to anticipate the consequences of development and ensure that undesirable changes do not take place.

A fall in water levels does not necessarily mean resources are being over-developed. When a well is pumped the water level declines as a gradient is created in the aquifer towards the well to allow water to flow towards it, and a cone of depression is formed; this occurs on a larger scale around a well-field. The size and shape of the cone depends upon several factors including the aquifer properties, the pumping rate, and the length of the pumping period (Fig. 10.8). There is need for concern when the abstraction rate exceeds the flow of water towards the well or well-field and water levels continue to decline for the same abstraction rate. The relationship between abstraction and water levels can be illustrated by considering the development of a confined aquifer.

As discussed above, when groundwater is pumped from a well in a confined aquifer the hydrostatic pressure declines and water is drawn initially from the vicinity of the well, as a result of a slight compression of the aquifer and a slight expansion of the water. As pumping continues the pressure is reduced over an increasing area until a recharge or discharge area is affected, when a new equilibrium is established by an increase in flow through the aquifer or a reduction in discharge from it. As the storage coefficient of a confined aquifer is small the effects of pumping expand rapidly.

In the London Basin, the Chalk and overlying Tertiary sands are confined by the London Clay (Anon 1972). Under natural conditions, part of the rainfall infiltrating the aquifers, where they crop out at the margin of the basin, flowed through the confined area to discharge in the Thames estuary. The aquifers have been pumped since the eighteenth century and water levels have fallen more than 70 m in some localities. Over considerable areas confined conditions no longer occur and appreciable volumes of the aquifers have been dewatered (Fig. 10.10). The change in water level due to pumping increased the flow from the outcrops, and also reduced and eventually stopped the discharge of water from the aquifer, the only outlet then being pumped wells. The natural flow through the central part of the confined basin was about 55 Ml/d but this increased, as a result of pumping, to a maximum of over 225 Ml/d in the late 1930s; in 1965 it was about 185 Ml/d.

The London Basin is often said to be over-developed because of

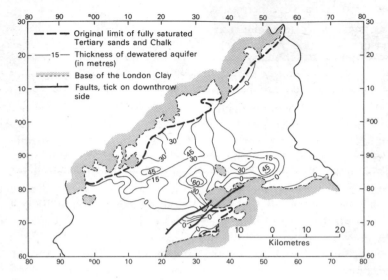

Fig. 10.10. Thickness of dewatered Tertiary sands and Chalk in the central part of the London Basin.

the large decline of water levels, but for each increment in the pumping rate a new equilibrium was established. During the periods when water levels were declining to new equilibrium levels, water was taken from storage but these were only transient phases. At each equilibrium stage related to a particular abstraction rate, water was derived by flow through the aquifer from the outcrop. Some 5.7×10^9 m³ have been pumped from the confined aquifers in the central part of the Basin in the 200 years since development commenced, with reduction in storage contributing 20 per cent and increased flow through the aquifer about 80 per cent.

Because of the much larger storage coefficient, pumping from an unconfined aquifer does not affect flows from the aquifer so rapidly as in a confined aquifer, but the discharge will nevertheless sooner or later be reduced by an amount equivalent to the volume pumped. The time required for this to take place depends upon the distance of the abstraction point from the outlet and the properties of the aquifer. It may take many years to reach a new equilibrium in an aquifer with a high storage coefficient and low permeability, even in small catchment areas, whereas a rapid response will be recorded for a very permeable aquifer with a low storage co-efficient.

In England, where river catchments are small, groundwater development has in many areas reached the stage where river flows are being significantly reduced, and this is particularly apparent and undesirable during the summer. In contrast, the southern High Plains of Texas and New Mexico are underlain by the Ogallala Formation, a Tertiary arenaceous deposit with an average thickness of 150 m. Natural recharge is small and the water resources are being mined, but the area is so large that pumping from the central regions of the Plains will not affect the natural discharge for centuries (Theis 1965).

Sea-water intrusion leading to the contamination of wells is a serious problem in coastal regions where groundwater is extensively used. Under natural conditions the hydraulic gradient is towards the coast and fresh water discharges to the sea, thereby preventing the encroachment of sea water. Because fresh water has a lower density than sea water, fresh water moves over the saline water and an interface, which slopes inland, exists between the two fluids forming the upper surface of a salt-water wedge (Fig 10.11). The position of the interface is approximately defined by the relationship that for every metre of fresh water above mean sea level there are 40 m below; this is the Ghyben–Herzberg relationship. When water levels are lowered in an aquifer near a coastline, the flow of fresh water to the sea is reduced and the interface rises; if the hydraulic gradient is actually reversed the rate of intrusion increases.

The classic example of the formation of a freshwater lens above saline water is in oceanic islands (Fig. 10.12). Under natural conditions the position of the groundwater level and hence the depth to the interface is a function of the amount of infiltration, the size of the island, and the permeability of the aquifer (Todd 1959). In such situations it is necessary to ensure that pumping does not lower the freshwater head to a level which allows saline water to rise into wells (Fig. 10.12b). Saline intrusion has been controlled at some locations in California by injecting water into the aquifer through a line of wells parallel to the coast to create a freshwater pressure ridge above mean sea level (Fig. 10.11c).

Appreciable land subsidence has often followed the development of groundwater from aquifers confined by relatively unconsolidated

Groundwater

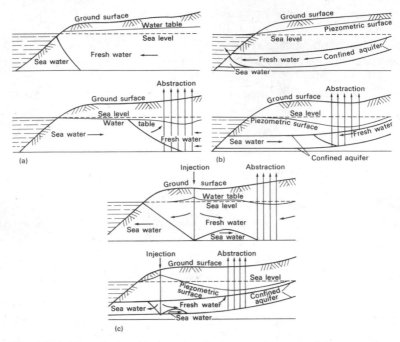

Fig. 10.11. The relationship between fresh groundwater and sea-water in a coastal aquifer: (a) an unconfined aquifer under natural conditions and subject to saline intrusion; (b) a confined aquifer in continuity with the sea under natural conditions and subject to saline intrusion; (c) the control of saline intrusion with a freshwater ridge in an unconfined and confined aquifer. (By courtesy of California Department of Water Resources.)

clays and silts, as in alluvial deposits. The reduction of artesian pressure allows water to drain from the confining layers with a consequent reduction in their thickness owing to consolidation. The amount of subsidence corresponds to the volume of water drained from the confining layer.

Subsidence exceeds 2 m over extensive areas in the San Joaquin Valley in California and maximum values are about 9 m. This is due to a reduction in artesian pressure of more than 50 m in alluvial deposits. Many other instances in similar geological situations are known including the Po valley in Italy, Mexico City, and Japan. In Tokyo subsidence of as much as 2 m has affected an area of more than 250 km² as a result of a decline in the piezometric surface by up to 80 m in a thick sequence of Quaternary deltaic deposits.

244

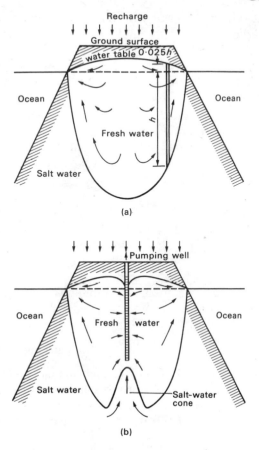

Fig. 10.12. Freshwater lens in an oceanic island under (a) natural conditions and (b) with a pumping well. (Reproduced from Todd (1959) by courtesy of John Wiley and Sons Inc).

5. Groundwater quality

The chemical nature of groundwater is mainly derived from the soil and rocks with which it is or has been in contact. Rainfall is a dilute chemical solution and it contributes significant proportions of some ions in groundwater, especially in regions with little soil cover where hard, compact rocks crop out. As water moves through the ground, solution continues and the concentration of groundwater tends to increase with the length of the flow-path.

245

Groundwater

The principal ions in groundwater are calcium, magnesium, sodium, bicarbonate, sulphate, and chloride. Many other minor constituents are present, some of which are important in specific fields; for example fluoride prevents dental caries when present in concentrations of about 1 mg/l, and nitrate is undesirable at concentrations in excess of 45 mg/l, since it causes methaemoglobinaemia in babies.

As rainwater infiltrates through the soil it comes in contact with the soil air, which contains 1 to 5 per cent carbon dioxide, considerably more than is in the atmosphere. The acidic solution formed reacts with carbonates in the rocks, giving solutions of calcium and, to a lesser extent, magnesium bicarbonates. This is a dominant reaction in limestones and in sandstones which are cemented by carbonates. If rocks, such as decalcified sands, do not contain carbonate but are overlain by a soil layer, the groundwater may be acidic and corrosive, owing to the carbon dioxide in solution, and the water is likely to be soft.

An important source of sulphate in groundwater results from the oxidation of iron sulphides, which are widely disseminated in sedimentary rocks. Sulphuric acid is a product of the reaction and this reacts with any carbonate present to form calcium or magnesium sulphates. In the absence of carbonates this oxidation produces a very troublesome iron-rich water commonly encountered during mining operations in argillaceous rocks.

Most of the chloride found in actively circulating groundwater is derived from rainfall although the chloride of most connate brines had its origin in the sea.

There is a general relationship between the chemical nature of a groundwater and climatic zones. Calcium bicarbonate waters tend to occur in temperate zones, whereas waters rich in chloride and sulphate are more usual in arid and semi-arid regions. Intensive leaching of the soil in the tropics gives a groundwater with low chloride and sulphate contents.

Solution reactions are generally dominant in aquifers at outcrop where water is actively circulating. In this situation the water in limestones and calcareous sands is relatively hard owing to the presence of calcium and magnesium bicarbonates and sulphates.

Water flowing through confined aquifers is commonly modified by ion exchange, the calcium and magnesium in the water being replaced by sodium from exchange media in the rocks. Sulphates may be reduced by bacteria, one of the by-products being carbon dioxide which is then available to dissolve calcium and magnesium carbonates in calcareous environments. Once in solution these cations are exchanged for sodium. High bicarbonate concentrations can also result from the oxidation of carbonaceous material if this is present in the aquifer. These various changes in a confined aquifer produce a soft sodium bicarbonate water, with possibly a low sulphate content, which contrasts with the hard water found in the outcrop zone (Fig. 10.13).

Groundwater is invariably saline in aquifers at depth where circulation is restricted, either because of low permeabilities or restricted outlets. Such waters are believed to have formed from the diagenesis of sea water trapped in sediments at the time of deposition, although the gradual solution of soluble constituents in rocks has also been advocated. During the course of time the

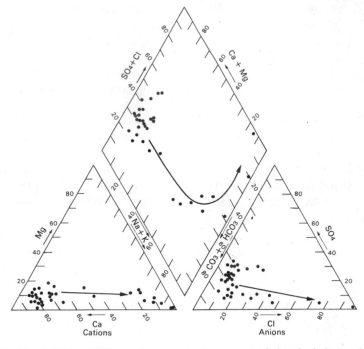

Fig. 10.13. Diagram illustrating a typical sequence of chemical change in a confined aquifer in a downgradient direction. The sequence shown by arrows is from a calcium bicarbonate water through a sodium bicarbonate water to a sodium chloride water in the Lincolnshire Limestone, England.

original interstitial water of sediments is redistributed by fluid potential gradients, including those induced by the consolidation of the sediments and later deposits. The high concentration of brines encountered at depth, often many times that of sea water, is due to rocks acting as semi-permeable membranes. Because of chemical reactions between the water and the rocks, the composition differs considerably from sea water. Commonly the Na/Cl, Mg/Cl, and SO_4/Cl ratios have decreased and the Ca/Cl ratio increased.

As groundwater flows from an outcrop to greater depths in a confined aquifer, the chloride concentration steadily increases as the percentage of connate water increases. The water eventually becomes non-potable (the accepted level for public water supply is 250 mg Cl/l) at varying distances from the outcrop depending upon the permeability and the position of any outlets from the confined aquifer.

To summarize, two sequences of chemical change can be identified in groundwaters. The first concerns meteoric water moving from an outcrop area to greater depths in a confined aquifer; the other is the modification of sea water once it becomes the interstitial pore water of sediments and is buried in sedimentary sequences. As a result of these processes waters at relatively shallow depths are usually bicarbonate waters; these grade into sulphate waters and finally into chloride waters. Saline waters moving through sedimentary sequences can dissolve and transport hydrocarbons and mineral ores (see Chapters 3 and 7).

Outcrops of permeable rocks are prone to pollution by leachates from landfills and mine waste tips and as a result of the disposal of sewage effluents, and agricultural and industrial wastes. Pollution of groundwater is often a slow, insiduous process, not readily apparent in the early stages; but once it has occurred, restoration to the natural state may take decades because of the slow movement of groundwater and its long residence time in the ground. For this reason anticipation of pollution hazards is essential so that precautions and preventative action can be taken.

The nature of an aquifer is obviously a principal factor determining the risk. Pollutants can move rapidly through a fissured aquifer, whereas unfissured arenaceous rocks exercise a considerable purification function by filtration and oxidation. Leachates from domestic refuse which have passed through about 10 m of sand are unlikely to contain significant bacterial pollution. The thickness required depends however, upon the nature of the deposit, and to ensure adequate bacterial purification water should flow through about 30 m of unfissured rock in the saturated zone; considerably less is necessary in the unsaturated zone. Chemical pollutants may travel much greater distances in the saturated zone before dilution reduces them to undetectable amounts but they may also be significantly reduced by ion exchange, adsorption, and precipitation.

Other factors important for assessing the pollution risk are the nature of the waste, the solubility of its constituents and their mobility in the ground, the thickness of the unsaturated zone (as this influences the extent to which pollutants are oxidized), and the direction of groundwater flow.

Groundwater pollution problems are commonly concerned with evaluating the velocity of solutes in water—in other words forecasting the rate at which a pollutant spreads through an aquifer. The velocity of water varies at different points in the intricate network of voids within an aquifer's matrix. The different flow rates of the various stream-lines results in the dispersion or spreading of the solute particles. Dispersion due to variations of velocity and the complexity of the voids in the aquifer is referred to as mechanical dispersion and it is mainly related to the geometry of the voids. Dispersion takes place in two directions: in the direction of flow it is called longitudinal dispersion; at right-angles to the flow it is lateral dispersion.

The spreading of solute particles is also increased by molecular diffusion caused by local concentration gradients of the solutes in the water. The consequences of mechanical dispersion and molecular diffusion are together referred to as hydrodynamic dispersion (Bear 1972). At low velocities molecular diffusion is important but as velocity increases mechanical dispersion increases in importance and eventually predominates. As a result of hydrodynamic dispersion a pollutant front in an aquifer is not sharp but is marked by a transition zone which increases in width with the distance travelled.

The movement of a solute may not be the same as that of the water molecules. Retardation can result from ion exchange and chemical reactions between the fluid and the aquifer. Different ions may be retarded by different amounts.

Problems causing concern at the present time are increasing nitrate concentrations owing to the disposal of sewage effluents into the ground and, in some intensively farmed regions, to the use of nitrogenous fertilizers; the potential hazards caused by the storage and transport of petroleum products; and the risk of contamination arising from urban run-off and leachates from solid wastes disposed of in landfills. It should be emphasized that groundwater pollution is commonly a local problem and potential hazards can often be identified. The most serious problem is basin-wide pollution caused by many widely dispersed sources, for example rising nitrate concentrations resulting from agricultural activity.

Increasing consideration has been given in recent years to the disposal of noxious industrial wastes into rocks containing saline waters at appreciable depths, which are isolated from freshwater aquifers by thick, relatively impermeable, argillaceous sequences. Many successful deep-well injection schemes have been carried out. Disposal of wastes into fractured crystalline rocks over 3500 m below the surface near Denver, Colorado, caused earthquakes, with epicentres at depths of 4·5 to 5·5 km, possibly due to hydraulic fracturing or a reduction in the effective stresses on potential fault-planes. This observation led to the suggestion that the build-up of dangerous stresses in rocks, which would eventually precipitate earthquakes, may be prevented by injecting water into fault-zones.

6. Groundwater management

Groundwater is unique as a mineral resource because it is continually replenished except in arid regions. The fact that it is renewable is always borne in mind when planning its use. In humid climates steps are usually taken to ensure that the total long-term abstraction does not exceed the average annual replenishment or, if it does, that the natural replenishment is supplemented by artificial recharge. Groundwater is, however, a resource that can lead to commercial wealth far exceeding the value of the water itself, and in some circumstances development in excess of average annual replenishment may be sound practice (at least initially) if the objects are to take advantage of the availability of a readily accessible cheap source of water and delay capital-intensive investment in more expensive water resource projects.

The advantage to be gained in the form of higher yields has led to the coordinated development of surface and groundwater resources, groundwater storage being used to balance both the short-term and long-term uneven temporal distribution of rainfall and hence surface run-off. With an increasing scale of groundwater development, there is a need to protect its quality and, if wastes have to be disposed of in, or on, an aquifer, to ensure that the factors controlling their movement through the ground are known. Groundwater basins may be simulated with mathematical models using digital computers or by electrical analogue models. The latter take advantage of the basic similarity between Ohm's law and Darcy's law. By using such models various development proposals can be examined and the most appropriate scheme selected. The movement of pollutants through an aquifer can also be studied and possible corrective methods assessed.

California is an example of an area where groundwater has been intensively developed, initially for irrigation but subsequently also for urban and industrial use. The alluvial lowlands of California have been formed by the infilling of low-lying depressed regions between mountains by sediments comprising sands, gravels, silts, and clays. In many of the individual alluvial basins, groundwater development exceeds the natural inflow; water levels have declined and the resource is being 'mined'. Methods adopted to correct this situation include the artificial recharge of surplus run-off, more efficient use of irrigation water, and the import of water by pipeline from areas farther afield. The aquifers are being used to store water at times of surplus and maintain adequate supplies during dry periods. Saline intrusion has occurred in coastal regions after the decline of water levels and, as mentioned above, steps are being taken to control this by artificial recharge.

The principal aquifers in England are the Chalk and the Triassic sandstones, which supply about 75 per cent of the total groundwater used. Most rivers in England are perennial and the development of groundwater has caused serious reductions in the dry-weather flows of many rivers supported by these aquifers. It is now increasingly accepted that aquifers provide not only for water supply but also for the maintenance of river flow for navigation, amenity purposes, and dilution of effluents. As in California, this has led to the integrated development of groundwater with surface water, again taking advantage of the large volumes of water stored underground. Water is abstracted from the lower reaches of rivers, which are regulated to a prescribed flow, by pumping groundwater into them as the need arises. In regions where river beds are relatively permeable the pumping regimes and the location of the wells are designed to have minimal effect on river flows by taking advantage of the time delay between cause and effect in unconfined aquifers. However, where river beds are impermeable, storage in the aquifer below the river can be developed. Pumping intercepts groundwater flow which would otherwise enter the river naturally but pilot studies have indicated that the net gain to river flow under favourable conditions exceeds 50 per cent of the pumping rate (Fig. 10.14). By pumping from the Chalk, the River Thames will be regulated in this manner to produce an additional yield of 450 Ml/d. It is expected that regulation of the River Great Ouse using some 345 wells will similarly provide up to 440 Ml/d. Schemes have been designed and tested in other areas, including proposals to regulate the rivers Severn and Yorkshire Ouse with groundwater from the Triassic sandstones.

Deltaic sediments represent a large proportion of the total volume of the world's sedimentary sequences and they therefore contain a significant volume of the groundwater resources, but development often creates problems associated with saline water, both surface and groundwater. Deltas are commonly very fertile areas and have considerable potential for economic growth. A case in point is the delta of the Rhine, Meuse, and Scheldt, one of the most intensively developed regions in the world.

The Netherlands occurs within this delta. Some 25 per cent of the country lies below sea level but along the coast extensive sand dunes have formed to heights of up to 30 m above sea level. These sands overlie Recent and Pleistocene deposits about 150 m thick and, under natural conditions, they contained fresh water in a state of dynamic equilibrium with the sea. This natural equilibrium was disturbed about a century ago when groundwater began to be developed to supply Amsterdam, the Hague, Leiden, and north-west Holland. Water levels declined several metres, and the saline interface rose, allowing brackish water to enter the deeper wells. The drainage of the sub-sea-level areas behind the dunes, known as polders, has also induced the inland flow of sea water beneath the freshwater lens in the dunes towards the artificial outlets created by drainage pumping. Since 1957 the fresh water storage in the dunes has been artificially replenished (Fig. 10.15) with water from the Rhine, which is pumped up to 80 km to the recharge areas. Water is recharged through basins with the recharge rate slightly exceeding the abstraction rate. Natural infiltration is supplemented by about 280 Ml/d.

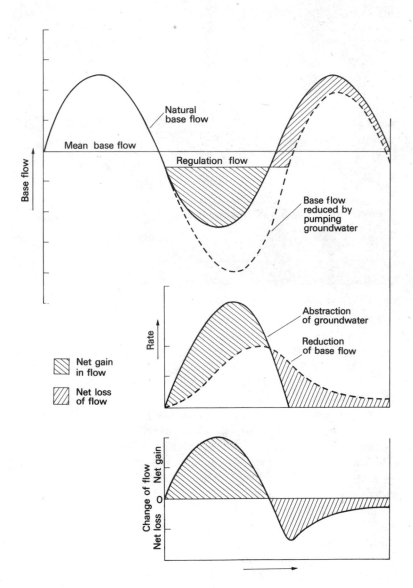

Fig. 10.14. Schematic diagram illustrating the regulation of river flow by pumping groundwater into a river.

Fig. 10.15. Recharge basins in coastal dunes, The Netherlands.

The Ruhr Valley in Germany contains an alluvial gravel aquifer 4 to 5 m thick with a typical hydraulic conductivity of 100 to 1000 m/d. Direct replenishment of the resource by rainfall is small in relation to the total volume of groundwater pumped. When the aquifer was initially developed, hydraulic gradients created by pumping induced the flow of river water into the aquifer. As demand increased the yield was supplemented by artificial recharge. The water is recharged through basins measuring some 250 m by 20 m and the average recharge rate is about 2·5 m/d. The water is recovered by pumping from wells and galleries after flowing about 50 m through the gravels. The city of Dortmund obtains about 100 million m³ of water a year by this method using 20 recharge basins with a total capacity of about 270 Ml/d, although the actual average recharge rate is only about two-thirds of this figure.

By constructing barrages and storage works to control river flow, irrigation of arid and semi-arid regions is possible for much longer periods than when a river system is not controlled. A consequence of perennial irrigation is, however, that downward seepage of water, in excess of that required by the crops, results in a gradual rise of the water table; and when it approaches the ground surface, evaporation leads to the deposition of salts at or near the soil surface. Under such circumstances a drainage system is necessary to control the water table. To avoid salination, irrigation water must be supplied in excess of that required by the crops and the water table must be controlled below the root zone. Any salt deposited in the soil zone will be leached by the excess water infiltrating through the zone.

Ancient civilizations dependent upon irrigated agriculture declined because of salt accumulation caused by poor drainage, and modern irrigation systems are also affected. Correction of the problem in the Lower Indus Valley in Pakistan will require many hundreds of wells to control the groundwater level. The well distribution and the pumping regimes have to be designed to avoid drawing underlying saline water into the wells in excessive amounts. (Bakiewicz *et al.* 1969).

The consumption of water in Israel exceeds the average annual natural replenishment. Ultimately this problem will be solved by an increase in the efficiency of water use, by treating effluents and brackish waters and by desalinating sea water, but at present the deficit is made up by mining groundwater resources. (Bonné *et al.* 1973). The two main aquifers in Israel are the Plio-Pleistocene sands and sandstones forming the Coastal Plain, which are some 200 m thick along the coast, and an aquifer of limestones and dolomites up to 1000 m thick, of Cenomanian and Turonian age, in the centre of Israel. Both aquifers have been intensively developed since the 1950s. Abstraction from the coastal plain in 1962 was two and a half times the sustained yield. Groundwater levels fell locally by up to 20 m and saline intrusion extended in places as far as 3 km inland. The importance of groundwater in the national economy and the need to protect the aquifer against contamination by the sea led to the careful control of the pumping regime. The regional water table was allowed to decline to reduce the natural outflow of groundwater to the sea, but when the desired groundwater level had been attained pumping was reduced to stabilize the extent of the rise and intrusion of the saline interface.

Since 1964 abstraction from both aquifers has been relatively constant. Artificial recharge of storm run-off and surplus winter surface run-off from the Kinneret basin has steadily increased, reaching 148 million m^3 in 1971. The water is recharged either by means of wells or by surface spreading. Near Tel Aviv saline intrusion is now controlled by artificial recharge.

Because of the alternation of permeable and impermeable deposits in sedimentary sequences and lateral changes in facies, together with the folding and tilting of such deposits after their formation, artesian conditions are relatively common. In artesian basins water may be abstracted many kilometres from the outcrop of an aquifer and the water stored in such basins is usually many thousands of years old. Even in the relatively small London Basin the water in the centre of the basin is over 20 000 years old (Smith *et al.* 1976). Carbon-14 measurements are used to determine the age of such waters. Some artesian basins are extremely large. Perhaps the best known is the Great Artesian Basin in Australia which covers about 1·5 million km², the aquifers being Triassic, Jurassic, and Cretaceous sandstones.

In north-east Africa an artesian basin referred to as the Nubian Artesian Basin (Himida 1970) underlies 2·5 million km² of Egypt, eastern Libya, the northern Sudan, and north-east Chad. Its boundaries include most of the Libyan Desert and a considerable part of the Arabian Desert east of the Nile Valley. The basin is formed in the Nubian Series, which ranges in age from Cambrian to Cretaceous. A continental facies of alternating sandstones and clays occurs in the south of the basin, passing laterally into limestones, dolomites, sandstones, and clays towards the north, as the thickness increases to some 3500 m along the Mediterranean coast.

The present recharge areas for the basin are the highlands of north-east Chad and northern Sudan, while the discharge is by upward leakage through confining beds in the north of the basin, including leakage to natural depressions in the Libyan Desert such as the Quattara Depression, the Siwa Oasis, and the Ghabub Oasis. The oases owe their existence to groundwater, and major land reclamation projects in their vicinities are based on groundwater development (Himida 1970). Although natural recharge is occurring at the present time where the aquifers crop out, the water in the confined region is 13 000 to 30 000 years old. Measurements of the stable isotopes, oxygen-18 and deuterium, imply that the water originated in a cooler period with a higher rainfall than that of the present day, that is during the Pleistocene (Munnich and Vogel 1962; Issar *et al.* 1972). As in many confined systems, the quality of the water deteriorates in a down-gradient direction and saline waters are encountered in the north of the Libyan Desert where the Nubian Series is overlain by younger rocks. Near the Mediterranean coast the water has a total ionic concentration of about 300 000 mg/l, mainly sodium and chloride ions. The close resemblance between the chemical composition of water in the Nubian Series along the Gulf of Suez and the hot brines in the deep basins of the Red Sea has led to the suggestion that the Red Sea brines, which are associated with sediments enriched in metals, are at least partly of connate origin (Issar *et al.* 1971).

7. Conclusion

The various groundwater developments briefly discussed in this chapter indicate the nature of the problems facing the hydrogeologist and the methods adopted to combat them. Since the 1940s there has been a considerable increase in the use of groundwater, but decision-makers considering large-scale water development alternatives have nevertheless often found groundwater less attractive than surface water. This is because a groundwater resource, as with any mineral, has to be proved and evaluated before development proposals have any true validity. Furthermore, the management of a groundwater reservoir is more complex than that of a surface reservoir. Data are required about variations in permeability and storage capacity, the sources of inflow, and the position of the various outlets, as well as the effect of pumping on both inflow and outflow. However, the principles of groundwater flow are now more fully understood and the technological ability to develop the resource is more widely available. These facts, together with the obvious cost–benefit advantages associated with groundwater development, including lower capital and running costs, less extensive land requirements, and the fact that development can be phased to meet the expansion of demand, should ensure a steady increase in the use of the large groundwater potential that exists in the world.

References

ANON. (1971). *Scientific framework of world water balance.* UNESCO, Paris.

ANON. (1972). *The hydrogeology of the London Basin.* Water Resources Board, Reading.

Groundwater

BAKIEWICZ, W., MILNE, D. M., and STONER, R. F. (1969). Groundwater development in the Lower Indus Plains, *Trans. Am. Inst. Min. Metall. Engrs*, **244**, 28–41.

BEAR, J. (1972). *Dynamics of fluids in porous media*. Elsevier, New York.

BONNE, J., CROSSMAN-PINES, S., and GRINWALD, Z. (eds) (1973). *Water in Israel—Part A Selected articles*. Ministry of Agriculture, Tel Aviv.

HIMIDA, I. H. (1970). The Nubian Artesian Basin, its regional hydrogeological aspects and palaeohydrogeological reconstruction. *J. Hydrol. (N.Z.)*, **9**, 89–116.

ISSAR, A., BEIN, A., and MICHAELI, A. (1972). On the ancient water of the Upper Nubian Sandstone aquifer in Central Sinai and Southern Israel. *J. Hydrol*, **17**, 353–74.

ISSAR, A., ROSENTHAL, E., ECKSTEIN, Y., and BOGOCH, R. (1971). Formation waters, hot springs and mineralization phenomena along the eastern shore of the Gulf of Suez. *Bull. int. Ass. scient. Hydrol.* **16**, 25–44.

MUNNICH, K. O., and VOGEL, J. L., (1962). Untersuchungen an Pluvialen Wassern der Ost—Sahara. *Geol. Rdsch.* **52**, 611–24.

PETERSON, F. L. (1972). Water development on tropic volcanic islands—type example: Hawaii. *Ground water.* **10**, 18–23.

SMITH, D. B., DOWNING, R. A., MONKHOUSE, R. A., OTLET, R. L., and PEARSON, F. J. (1976). The age of groundwater in the Chalk of the London Basin. *Wat. Resour. Res.* **12**, 392–404.

THEIS, C. V. (1965). Ground water in south-western region, 327–41 in Fluids in subsurface environments. *Mem. Am. Ass. Petrol. Geol.* **4**.

TODD, D. K. (1959). *Ground water hydrology*. John Wiley, New York.

11. Geology in the construction industry

1. Introduction

The engineering geologist working in the construction industry is a member of a team having responsibilities for conceiving, planning, and bringing into being capital projects such as multipurpose water resource schemes involving dams, reservoirs, tunnels, and power stations: communication networks involving motorways and bridges: and urban and industrial developments involving structures on all scales, together with the associated services. It is the responsibility of the engineering geologist to ensure that the geological conditions, in the widest sense, in which any structures are to be located are predicted correctly so that they may be constructed within estimates of time and cost and can subsequently operate safely and effectively. In addition, it can be necessary for the engineering geologist to provide geological information related to other fields such as the availability of groundwater, sources of construction materials, natural hazards, and the environmental impact of the proposed engineering development.

Irrespective of whether the scheme is a major project costing several £100m or is a small two-storey housing development, the same basic factors will determine the manner in which geological expertise is applied to the realization of the engineering objectives. First, it is important to appreciate that this realization can be achieved only after a series of well-defined steps; secondly, to put the scheme into effect requires satisfactory collaboration between client, consulting engineer, and contractor (Table 11.1). The client is, fundamentally, the customer and he will decide at some stage that there is a need for a specific development. In situations determined by population growth there can be a continual forward review of needs against projected demands so that adequate supplies of energy, water, and housing, for example, can be available when required. In other situations, such as in response to a rapid growth of an extractive industry, the need may develop quickly and may possibly outstrip the available resources or technology. Once the client has adequately outlined the need it is common practice for the consultant to be asked to define the need more precisely, examine alternative solutions and identify a preferred solution, carry out a detailed investigation and design the proposed solution, develop these designs to the extent that construction can be carried out, advise on the promotion of the scheme and then on tenders for construction obtained from contractors once approval for construction has been obtained, supervise construction, and, finally, monitor performance of the completed scheme. The contribution made by the contractor is to provide services for site investigation, by drilling and other means, and for the physical construction of the scheme.

TABLE 11.1

Phases of development of an engineering scheme

	Consulting engineer	*Contractor*
Investigation Design	Desk study	
	Feasibility report	Site investigation/testing
	Project report	services
(Promotion)	Contract documents	Tender evaluation
	CONTRACT AWARD	
Construction	Supervision and design variation	Construction and variations leading to claims
	COMPLETION	
Operation	Observation/recording	Maintenance
	Design of remedial works	Construction of remedial works

Increasingly, particularly in the case of national or regional authorities, the task of investigation, design, and supervision of construction is totally or partly carried out by the authority. It is also not unusual for the larger authorities to have their own construction branch so that the complete responsibility for all aspects of the scheme is taken by the authority; in such circumstances the separate roles of client, consultant, and contractor can become fused into one. Apart from this situation, clients are more generally taking an active role in the essentially engineering aspects of a scheme, partly because of the implications of inflation on costs and the need for forward budgeting.

Engineering geologists find employment with clients, consultants, and contractors and become involved in all phases of an engineering scheme from initial evaluation of the alternatives to the monitoring of the operation of a completed scheme. However, the engineering geologists' roles during these phases will differ as the engineering requirements progressively change and develop.

2. Investigation

2.1. Investigation phasing

The initial phase of investigation will consist of a review of existing information, such as published maps, reports, and air photographs, supplemented by a 'walk-over' survey of the study area. Although no new information will be collected during this phase, it will lead to a more precise definition of alternatives and the elimination of unsuitable sites. An important function of such a desk study will be the planning of the investigation needs for the scheme, both in terms of technique and scale. The subsequent investigation will generally be divided into three specific stages.

Stage 1. The primary objective of this stage will be the definition, in as precise terms as possible, of the location of the various parts of the scheme. In general, limited exploration will be carried out and this will have the objective of defining the main factors which will influence feasibility, cost or safety. By the close of Stage 1, the basic practicability of the project should have been established.

Stage 2. During this stage, the foundation for each structure or each route will have been explored and laboratory testing carried out. If the conditions prove less favourable than predicted, and alternatives are available, then the programme of investigation may be adjusted. Sources of construction materials would be located and wider-scale investigations (e.g. reservoir watertightness) carried out.

Stage 3. This stage will be concerned with studies related to detailed design, possibly involving closely-spaced drilling, adits, shafts and testing *in situ*. The sources of construction materials will be evaluated in both quantitative and qualitative terms.

Foundation treatment trials and other prototype experiments will be carried out.

In practice, these phases may overlap with one another or there may be sound engineering reasons for separating one phase from another. However, it is necessary to ensure that each phase, however executed, is designed in such a way as to provide answers to engineering questions which will inevitably be asked. For example, in the case of a concrete dam it will be necessary to establish the following: probable foundation depth; implications of the strength and deformability of the rock below the foundation level on the dam behaviour; relative watertightness of the rock foundations, existing groundwater conditions and consequential inflows; technique of excavation and likely fragmentation of excavated rock; stability of side-slopes of excavations; existence and distribution of significant defects in, near, and below the dam foundations; and engineering techniques required to ensure that the quality of the foundations meets an acceptable standard. The investigation must be phased and planned in such a way as to ensure that the information necessary to answer the requirements is obtained, and in many cases the geologist has to anticipate those requirements at a stage when only limited geological information may be available. In general terms, the investigation will move progressively from a phase of essentially geologically orientated studies (Stage 2) to a phase when the engineering implications of the site geology are determined. There must therefore be close cooperation between the engineering geologist and the design engineer in establishing and monitoring a programme of investigation. For this reason it is important to establish, within any such investigation, a statement of the geological and engineering objectives of each component of the programme. By this means it is easier to recognize unanticipated results and change the direction of the exploration programme.

2.2. Investigation methods

For most civil engineering works it is unusual for investigations to extend deeper than about 200 m. For this reason, the drilling or geophysical techniques used are not uncommonly scaled-down versions of those applied in the mineral or hydrocarbon extraction industries. Investigations can be required to greater depths in connection with tunnels, underground chambers, deep foundation studies for large dams, or reservoir watertightness investigations.

Geological mapping. Surface geological studies by systematic mapping are an essential preliminary to all engineering investigations. In urban areas, where natural exposures are absent or limited, it may be necessary to obtain data from other investigations or to examine construction records in order to supplement published maps. Elsewhere, it would be common practice to re-map the area, supplementing the mapping from existing sources and by the use of air photographs which may already be available or could be specially flown; a wide variety of remote sensing systems, such as satellite, colour, false-colour, and infrared can be used. There are differences between conventional geological mapping and that used for engineering purposes which can be described as engineering geology or geotechnical mapping. The fundamental difference in approach would be that a geological map is essentially a record of surface geology with some subsurface extrapolation in areas of mining. However, for engineering purposes, it is necessary to obtain information such as that concerning slope stability, groundwater discharge, areas of subsidence, construction materials, and detailed rock-mass structure. Inevitability the level of detail required for construction purposes is much greater than that needed for a broader view of regional geology. The mapping scales used would preferably be as follows.

1:50 000: for the complete project area extending perhaps 5–10 km beyond its limits;
1:5000: for all individual parts of the project such as tunnel lines and reservoir sites;
1:1000: for all structures such as dams, bridges, and power stations.

It is of particular importance to appreciate the regional setting within which a project lies. For example, a multistorey block to be located in an area of the Coal Measures may self-evidently have foundations influenced by previous coal extraction. The regional implications may, however, be less obvious but no less significant. During the early phases of construction of the Mangla Dam in Pakistan, shear zones parallel to the gently dipping bedding were identified and, in view of the implications for the strength of the foundations, the design was adjusted. These shear zones appear to have been the consequence of bedding-plane slip associated with concentric folding. It was then appreciated that the presence of a monoclinal fold some distance downstream of the dam could have been a contributory factor in the generation of these shear zones. The progress of the geological mapping will normally continue throughout the site investigation, being progressively reviewed in the light of new information. It is important to ensure that at an early stage adequate geological information is available to ensure that the appropriate site investigation methods are identified so that they may be applied subsequently to the more detailed studies.

Engineering geophysics. The two main methods of geophysical exploration used in engineering are the seismic and resistivity techniques. Seismic methods are particularly suited to solving problems of determining depth to rockhead (bedrock surface) or depth of weathering, which commonly arise in designing the foundations for buildings. Equipment particularly suited to the relatively shallow exploration depths required has been developed for the engineering industry and this commonly operates on a dropping weight or hammer blow as the energy source in order to reduce the inconvenience of handling explosives, and the associated risks. The measured longitudinal wave velocity (P-wave) is a function of the density and modulus of dynamic deformability of the rocks tested and can, therefore, be used as an indication of rock quality. Sound, massive granite below the water table will have a P-wave velocity of 4500 m/s or more, whereas more fractured, altered or less competent rock will have a proportionally lower velocity. By this means, it is possible to obtain a prediction of the relative deformability of the foundation or its response to different methods of excavation. Resistivity methods are also used to determine rockhead, although are generally less accurate than seismic methods, and have also been applied where there are rapid changes in the level of rockhead, such as over collapsed limestone caverns. However, resistivity methods have proved most successful in the identification of the position of the water table in what are essentially overburden materials such as sands and gravels. Changes in gravity are not normally great enough to be detected even by sensitive methods but they have been used to detect rockhead levels in basic gneiss, caverns in limestone, and buried topography in arid zones for groundwater studies. Magnetic techniques can be applied to situations where there are relative changes in rock magnetism appropriate to the engineering problem under study, and also in the location of shallow mine workings, particularly shafts. Geophysical methods can rarely provide enough positive information for a total engineering solution but, supplemented by other techniques such as mapping and drilling, they can both reduce the cost of exploration and improve the level of interpretation.

Drilling. As with other branches of industrial geology, drilling provides one of the most positive methods of establishing the actual geological conditions *in situ.* Apart from enabling the subsurface geology to be interpreted, drilling also provides the means of obtaining undisturbed samples for subsequent testing, carrying out small scale tests *in situ*, and instrumentation; for the purposes of economy most boreholes are designed for multiple purposes. Techniques of drilling have been devised for overburden, residual, and drift materials (commonly referred to in engineering practice as 'soils') by soft ground methods and, in part separately, for rock materials. The soft ground methods cover a range of separate techniques which can recover samples in either a disturbed or an undisturbed state. Disturbed sampling methods include hand augers, power augers, mounted continuous flight augers, and various forms of percussive boring where the sample is recovered by means of a bailer. The undisturbed sampling methods include drive sampling (an extension of the percussive technique), continuous tube sampling, and various adaptations of rotary drilling methods used for rocks. One of the commonest systems of soft ground boring used in Britain is the shell and auger rig (Fig. 11.1). For many engineering purposes where the hole is required for testing purposes or for instrumentation, it may not be necessary to obtain undisturbed samples. Where, however, the materials are required for testing purposes it is essential that the samples should be recovered in as undisturbed a condition as possible. In real terms, it is impossible to obtain a truly undisturbed sample since the confining stress and groundwater conditions are inevitably modified as part of the act of sampling. The type of material may be such as to make sampling impractical, for example in the case of openwork gravels, bouldery or gravelly tills, and water-bearing silts, all of which can be extremely difficult to sample *in situ*. For this reason, various forms of field test have been devised which can measure the

Fig. 11.1. Shell and auger rig for soft ground boring.

permeability or strength of the soil materials forming the walls of the borehole. Although percussive methods, entailing the recovery of chippings, are used for special purposes, the main method of rock drilling uses a rotary drill, a double- or triple-tube core barrel, a diamond bit, and water, air, or mud as a flushing and lubricating medium to aid cutting. The type and diameter of barrel and bit and the flush method are selected depending on the ground conditions and availability of equipment. Large-scale calyx drilling can be used to recover massive cores (Fig. 11.2) and the large-diameter hole then provides a small shaft which can be directly inspected *in situ* (Fig. 11.3). The object of rock drilling is to obtain full recovery of all the rock penetrated, and improved recovery can be achieved

Fig. 11.2. Large diameter core of Hawkesbury Sandstone from Warragamba Dam, New South Wales.

by changes in the method of drilling used. However, it is inevitably the most fractured and altered rock which is the most difficult to recover and the most relevant to most engineering problems. Probably the single most important task performed by the engineering geologist is the recording of geological conditions in rock cores (Figs. 5.8 and and 6.5) and their engineering interpretation (Anon. 1970) in relation to the particular engineering project. The degree of fragmentation of the core is a useful measure of rock quality; the more closely spaced fractures are associated with shorter lengths of core, or fragmental rock, and thus poorer rock conditions. That proportion of the core recovered in sticks

267

Fig. 11.3. Engineering geologist about to be lowered down large-diameter borehole in Chalk on route of M 40 motorway.

longer than 100 mm is referred to as the rock quality designation (RQD); this index is used extensively as an approximate guide to the potential engineering behaviour of rock masses (Table 11.2). Before drilling a borehole in soil or rock it will be necessary to decide on a programme of sampling, testing, *in situ*, and instrumentation. Valuable information can be obtained from the behaviour of groundwater levels on a day-to-day basis during drilling, and this may be used to locate piezometers (standpipes sealed-in over certain levels to determine the effective groundwater head). Groundwater, in particular, has a considerable influence on the behaviour of soils and rocks, and it is essential that adequate information is obtained during the investigation, prior to design and construction, as to the position of the water table and the distribution of head with depth together with related seasonal variations.

TABLE 11.2
Rock quality

RQD (%)	Fracture frequency (Number per metre)	Rock mass description
0–25	> 15	Very poor
25–50	8–15	Poor
50–75	5–8	Fair
75–90	1–5	Good
90–100	< 1	Excellent

Excavation. Excavation enables the geological conditions to be determined less ambiguously than is commonly the case with information derived from borehole cores. In addition to detailed recording it is also possible to obtain large undisturbed block samples for laboratory testing and there is adequate space for large-scale tests *in situ*. The methods of excavation include trial pits, shafts, trenches, adits, and tunnels. The development of hydraulic excavating equipment has resulted in greater economy and speed in the formation of trial pits or trenches, which can now readily be excavated in soils and in soft or weathered rocks to depths of 5 m or more. With such temporary excavations there is some risk of face collapse unless support is installed; in any event, the depth to which such excavations can be carried out will depend on the location of the water table, unless pumping equipment is provided. In some countries or in remote locations it is possible that the cost of excavation may be less than that for drilling to yield equivalent geological information. In these circumstances there may be financial advantage in using shafts or adits in place of boreholes. Most engineers prefer to be able to observe the condition of a rock mass directly *in situ* rather than relying on the interpretation of borehole cores; it is also desirable for exposures to be available for inspection by a contractor at the time of tendering for construction.

Samping and testing. An important difference between engineering geology and other branches of applied geology is that engineering geology is concerned to a significant degree with material properties and their response to different environmental conditions. One of the primary purposes of a site investigation is to provide soil, rock, water, and gas samples for subsequent laboratory testing, or to provide access for testing *in situ*. The main properties of soils and rocks which are of engineering significance are their strength, deformability, permeability, and chemical stability. Samples collected in the field may be in the form of disturbed samples (generally contained in bags and containers) or undisturbed tube or core samples and block samples. Undisturbed samples are sealed to maintain their natural moisture content. All samples are generally subjected to classification tests which provide a broad definition of their engineering characteristics; such tests should include, for soils, grain-size analysis and liquid/plastic limit; for rocks, porosity and slake durability. Prepared undisturbed samples are subjected to more elaborate tests which will provide data relevant to engineering design concerning deformation and failure under load, and response to groundwater flow. An equivalent range of engineering tests has been developed, to varying degrees of sophistication, for use under field conditions. For example, cylindrical expanding jacks installed in boreholes can be used to determine deformability moduli, and large blocks can be excavated and then loaded until shear failure occurs. If long-term ground movements (e.g. a failing slope, subsidence), or changes in groundwater conditions are expected, appropriate instrumentation can be installed.

2.3. Design of investigation

The size and form of the investigation will depend on a wide variety of factors, including the type of project, degree of access, climate, financial restraints, and degree of urgency as well as the site geology. It is therefore necessary to design the investigation within these restraints, tailor-making it to the particular circumstances, and it will only be in particular cases (e.g. urban developments in an area underlain by uniform geology) that the planning of site investigations may tend towards a uniform pattern. In the case of a relatively simple multistorey block to be founded on horizontally bedded strong sandstones with thin overburden, little more than two or three boreholes, a piezometer, and some simple rock-strength tests might be required. However, if the scheme involved several multistorey blocks then it could be economical to carry out a field plate bearing test to measure the deformability of the sandstone, thereby achieving economies in the multiple foundations. A similar multistorey block might be founded over dipping Coal Measures and directly underlain by glacial drift. In this case, several boreholes might be required to determine whether the foundations should be in, or on, the drift or the Coal Measures. A range of potential problems can be envisaged: deformable clay layers or water-bearing sands in the drift, a variable level of rockhead, old mine workings in the bedrock, and so forth. A number of boreholes (drilled by different means) and a programme of laboratory tests and field instrumentation would be required to obtain adequate information. Moving forward to an entirely different scale of investigation a large project may require a year or more for complete investigation, requiring the creation of access, site camps, and essential services. The planning of complex site investigations requires special skills because it is necessary to anticipate the conditions (as yet unrevealed) which it may prove necessary to explore. In tendering for the contract, the site investigation contractor will base his rates on the requirements of the client or consultant as set out in a Specification and Bill of Quantities. If the likelihood of success of exploration or sampling with particular methods (e.g. geophysics) is in doubt then such situations can be covered by the definition of provisional items in the Bill to be developed if required.

2.4. Role of the engineering geologist in investigation

The engineering geologist, employed by client or consultant, will be closely involved in the detailed planning of the investigation, supervision of work in progress, and interpretation of the results (Edwards 1972). He may be resident on site, taking responsibility not only for recording of geological information but also selection of sampling, checking tests *in situ*, and locating instruments. The role of the engineering geologist employed by a site investigation contractor will be similar except in that he will take a greater part of the management of the day-to-day functions of drilling, sampling, and testing, commonly adopting a site supervisory role.

3. Design

3.1. Introductory

The object of an investigation is to generate adequate information upon which the scheme or structure can be designed and costed. Inevitably, therefore, the design phase overlaps with that of investigation; as data are obtained from the exploration so there is a progressive development of the design and feedback of additional requirements from the investigation. This necessitates a considerable degree of rapport between design engineer and engineering geologist, and the design stage is probably the period when the geological contribution to the project is of greatest importance. The primary requirement is to ensure that the engineering proposals are compatible with, and can be adjusted to, the geology. If the site presents particular hazards, then the scheme must be relocated or modified in such a way as to cope with these hazards. In a region of active seismicity, for example, it is necessary to design structures to withstand the ground displacements consequent on an earthquake of the maximum probable intensity. If, however, there is a known, active fault attempts will be made to avoid locating structures above or close to it. For example, in an aqueduct on the California State Water Project advantage has been taken of a sag pond along the line of an active fault by discharging water from the aqueduct into one side of the lake and abstracting it from the other side of

Fig. 11.4. Asphaltic protection to limestone ridge on flank of Mornos Reservoir, Greece.

the lake; by this means the lake provides a temporary reservoir storage, no structure crosses the fault line, and there is surface access to the aqueduct in case of accident. Similar questions arise in the assessment of reservoir feasibility because a large capital expenditure must be invested in the project before the watertightness of the reservoir basin can be proved by impounding. It is necessary to consider the geological and pre-existing groundwater conditions during the investigation in order to establish whether there is, or could be, groundwater flow out of the basin. If potential leakage paths are detected it may be necessary to select an alternative site or carry out appropriate measures to reduce or prevent water loss. In the case of the Mornos Reservoir, west of Athens, a limestone massif in the reservoir is in direct contact with the sea some 20 km to the south. The depressed groundwater levels in the limestone suggested that seepage could occur from the reservoir into the limestone. In consequence, the exposed face of the limestone hillside which will be in contact with the reservoir water has been sealed by an asphaltic layer (Fig. 11.4).

The role of design is inevitably dependent upon the type of engineering structure involved. Three examples will illustrate the implications of geology for design.

3.2. *Foundations of Hartlepool Nuclear Power Station*

The foundations of buildings can be supported by both overburden and rock materials, the type of foundation being determined by the depth of overburden, the properties of the soil and rock materials, and the structural loading. Soils and rocks deform under load and this results in settlement. If the settlement is at a level which does not cause structural distress by cracking in the building, then the foundation loading has not exceeded the allowable bearing pressure. If the allowable bearing pressure is exceeded, then deformation will occur in the building leading to cracking and distortion of the walls and floors. At a high level of loading, failure will occur in the foundations if the ultimate bearing pressure is exceeded; such failure is more characteristic of structures built on soft clays, or influenced by seismic loading. If the near-surface materials have an acceptable allowable bearing pressure, then the foundations can be composed of spread footings. However, if there is a thick cover of relatively soft overburden the foundation loading can be spread over a raft at shallow depth or carried to depths by piles drilled or driven through the overburden to a firmer stratum (Fig. 11.5).

The Hartlepool Nuclear Power Station is located on the north-east coast of England and was constructed during the 1970s. The site is underlain by some 7 m of soft esturaine alluvium overlying a glacial complex ranging from 25 to 40 m in thickness; in consequence rockhead varies in elevation by about 15 m below the site. The glacial complex includes layers of stiff till and more complex water-bearing sands, gravels, and silts. The requirements for the foundations of a nuclear power station, particularly the reactor

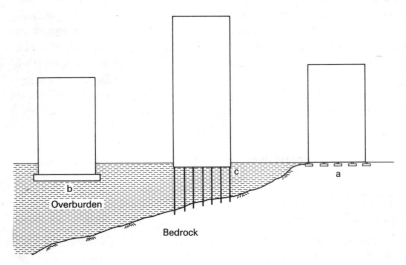

Fig. 11.5. Main foundation types: (a) spread footings, (b) raft, (c) piles.

block, are very stringent because of the high, concentrated loadings involved which do not permit significant differential settlement. The soft alluvial cover was unsatisfactory as a foundation material and was, in any event, to be removed in those areas where structures were to be built by bulk excavation. There were three possible foundation solutions to be considered: a raft founded on top of the glacial complex; driven or bored piles extending into the till layers; or piles carried down into the Bunter Sandstone bedrock. In view of the uncertainty associated with possible deformations in the glacial complex and the high loadings involved, it was decided to found the reactor block on a group of concrete shafts 2·3 m in diameter carried down into the sandstone and penetrating through the upper more weathered rock. For the turbine hall, where the foundation loadings are not so high and the criteria not as stringent, it proved possible to use a pattern of driven cast *in situ* concrete piles carried down into the main sand and gravel layer.

The decision to adopt a particular solution must inevitably be based on a large number of considerations. Apart from the factors considered in the previous paragraph, other matters considered included the groundwater conditions in the overburden and bedrock, a need to ensure that any piled foundations were not set too close together, and an appreciation of the physical problems encountered in attempting to excavate an exploratory shaft to bedrock, during the investigation, in which it had been planned to carry out loading tests on the rock *in situ*. As a consequence the evaluation of the properties of the bedrock was carried out using a combination of borehole interpretation, the results of a detailed cross-hole velocity survey, and a programme of laboratory testing on rock cores.

3.3. Farahnaz Pahlavi buttress dam, Iran

The Farahnaz Pahlavi dam, which has a maximum height of 112 m,

Fig. 11.6. Farahnaz Pahlavi Dam, Iran.

was constructed during the early 1960s on the Jaje-rud east of Tehran (Fig. 11.6). Upstream of the dam the valley broadens out to define the reservoir basin, one flank being composed of poorly cemented sandy gravels. Downstream of the dam, the river flows in a gorge some 2 km in length mantled by superficial and slipped materials on one flank. The dam is located at the only point in that gorge where stable bedrock occurs on both sides of the river. In general, dams are located at constrictions in rivers where a dam can be constructed at minimum height and volume; upstream the reservoir basin should broaden out into a level area, so providing maximum storage at minimum cost. At Farahnaz Pahlavi, the location of the dam was determined by the engineering requirement to store a specific quantity of water and the geological requirement of continuous bedrock to form a relatively watertight and stable foundation to the dam.

In the early stages of investigation consideration was given to more than one type of dam at the site. One possibility was the construction of an embankment dam which could have been constructed either of rock-fill or of alluvial materials, both of which were available locally. However, sources of low-permeability materials, essential for the core, were not obvious and the scale of the dam would have been such as to overlap on to slipped ground. The alternative was a concrete dam of either mass gravity or buttress type. A gravity dam is less economical in the use of concrete and would not provide the necessary internal flexibility in the dam that was required by the rock foundations. The buttress design minimizes the use of concrete and permits differential movements between the separate buttresses, and this design was selected. The selection of this type of dam nevertheless posed a number of geological problems.

The foundations of the dam are composed of quartzitic sandstones, shales, and dolomitic limestones, the last providing a source of aggregate. The left bank of the dam is founded on a spur of closely folded sandstones; downstream of this spur (Fig. 11.6) the rockhead is buried by about 35 m of overburden so that the dam had to be founded on a narrow ridge of rock with little effective support in a downstream direction. Stability was achieved by founding the buttresses at adequate depth in the ridge, well over 30 m below the level of the crest of the spur. In view of the relatively poor rock conditions in the spur (the borehole cores, for example, had low values of RQD) the buttresses were provided with a continuous plinth which was partially reinforced by post-tensioned cables, the foundations were stiffened by injection of cement–water grout, and an extensive drainage system was installed below the dam foundations in order to minimize uplift pressures. The right bank was composed of a steeply sloping face underlain by layers of sandstone and shale dipping in an upstream direction (Knill and Jones 1965). This created special problems with regard to both the relative deformability of the rocks and their strength; a passive thrust block was provided downstream of the dam against which the buttresses could transfer load.

3.4. Kielder tunnels

The investigations carried out for tunnels need to establish information in relation to the following: support measures required during construction, and subsequently, method of excavation, likely groundwater inflows, and any particular hazards (e.g. gas, high temperatures) or problems (e.g. exceptionally hard rock, swelling rocks). However, in view of the depth of cover and length over which investigations have to be carried out, the information available before construction is more limited than in other fields. For large-diameter tunnels, particularly those constructed under water, it is normal practice to excavate a pilot tunnel in advance of the main tunnel; the pilot tunnel is used both for investigation (thus proving the ground) and for subsequent ground treatment by grouting and other means. Nevertheless, it is still necessary to predict conditions ahead of tunnelling, and for this purpose considerable reliance is placed on geological mapping and consequent interpretation.

The Kielder tunnels total some 32 km and form an aqueduct system in north-east England to carry water south from the River Tyne to the Rivers Wear and Tees to the south (Carter and Mills 1976). The tunnels cross a gently dipping sequence of Carboniferous rocks including shales and mudstones, limestones, and sandstones. The investigations were directed towards establishing the geological sequence and structure by mapping, supplemented by a limited number of deep boreholes. By such means it was possible to draw sections along the tunnel lines, thus indicating the proportions of different kinds of rocks which were likely to be encountered, together with the locations of faults. Apart from providing samples for laboratory testing, the borehole cores provided an indication of the rock structure. Most reliance on information related to rock structure had, however, to be placed upon surface studies of natural exposures and quarries, account being taken of the weathering effects which have limited effect on the deeply buried rocks likely to be encountered.

The selection of support measures will depend on the rock-type and the associated mass structure, including the joint distribution. The more massive, blocky rocks will tend to be self-supporting, individual beds forming beams spanning the tunnel roof; falls may tend to occur when joints are clay-covered. In the shales and mudstones, disintegration of the rock is likely to take place, requiring the installation of full support soon after initial excavation. In order to select the form of temporary support required, trial adits were constructed and alternative support measures installed and then instrumented to determine the yield in the supports and the surrounding rock. Ideally no support is required but this can usually be achieved only in massive rocks. A variety of support measures can be provided based traditionally on the use of steel arches. In the simplest case such arches are installed with intervening lagging to prevent the collapse of small fragments; if lateral loads are imposed on the tunnel, invert struts are placed between the feet of the arches to prevent inward movement. Swelling ground conditions may require the use of circular steel ribs, the intervening spaces being infilled by steel lagging or concrete (Fig. 11.7). An alternative technique is based upon the use of sprayed concrete, installed as soon after excavation as practicable, supplemented by netting reinforcement and rock anchors installed in drilled boreholes.

A further question which arose relates to the method of excavation. Such rocks would commonly be excavated by drilling and blasting. In view of the uniformity of the rock materials and their strength, the possibility of a continuous tunnelling machine had to be considered. The maximum strength of rocks which can be cut with a full-face tunnelling machine is of the order of 200 MN/m² and only the limestones approach this strength. An additional problem which arises in using such a machine is that it fits so tightly into the excavated tunnel that support cannot be installed until the machine has moved forward to expose the roof. In rocks such as shales or mudstones the rate of disintegration can be such as to

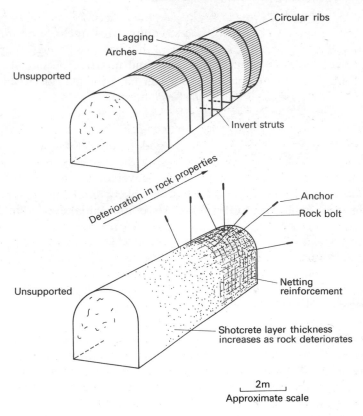

Fig. 11.7. Forms of tunnel support.

inconvenience the tunnelling operations well before support is in place. Quite apart from the engineering questions of excavation and support, hazards such as groundwater inflows and methane occurrences have to be predicted ahead of tunnel advance.

The most economical tunnel route is a straight line and it would be relatively unusual for a tunnel to deviate from such a route for geological reasons alone; the cost of additional tunnelling not uncommonly outweighs the probable cost of overcoming the anticipated geological hazard. Tunnel portals must be installed in stable ground, particularly since there will be a concentration of men, materials, and plant close to the portal for many years during construction (Fig. 11.8).

279

Fig. 11.8. Portal of Kielder tunnel in Wear valley with cutting in sandstone overlain by glacial drift.

3.5. Contract documents

The product of the design phase is the preparation of contract documents (Table 11.1) embodying the results of the design procedure in a form which will enable precise costing by a tendering contractor.

4. Construction

4.1. Role of the engineering geologist

Provided the investigation has been carried out properly and the geological observations have been correctly taken into account in the design procedure there should be no need for geological activities during construction. It is, however, unlikely that such reliance can be placed upon a site investigation unless the geological situation or engineering requirements are straightforward. In the general case it is important to ensure that the assumptions made during design, and therefore the geological conditions encountered, are as originally anticipated. The primary role of the engineering geologist during construction is to identify such changes in conditions from those predicted so that designs can be modified

Fig. 11.9. Engineering geologist inspecting foundations of a buttress, Roseires Dam, Sudan.

appropriately and construction procedures adjusted and, also, to confirm that the geological conditions are consistent with the design assumptions. Where hazards, such as water inflows, landslides, and tunnel instability, are expected during construction it will be necessary to take precautions in order to minimize risk and additional expenditure. Systematic inspection and recording of excavations will be essential before concreting or covering by fill in order to confirm and then record the anticipated conditions (Fig. 11.9).

In certain cases, the investigation carried out prior to construction is necessarily limited. This limitation influences the approach to construction. Under these circumstances it is recognized that design adjustments will have to take place during construction, and appropriate allowances can be made in specific circumstances. In the construction of large embankment dams it may not be economic to explore borrow pits or quarries in detail once it has been established that there is adequate material available. Once extraction starts, the design can be adjusted by placing material of particular characteristics in different parts of the structure; by this means more effective use can be made of the available plant on the

Fig. 11.10. General view of Rodovia dos Imigrantes motorway, Brazil.

site. If access to the site has proved difficult, then little information may be available before construction. For example, the Rodovia dos Imigrantes motorway was constructed over a section of the Serra do Mar coastal escarpment between Sao Paulos and Santos in Brazil. The escarpment is composed of metamorphic and granitic rocks weathered to a depth of about 40 m; removal of the cover of vegetation exposes the weathered rock and this results in instability. The escarpment slopes at 35° to 40° and is indented by steep gullies covered in thick jungle; access was virtually impossible before the construction of access roads, which are probably double or treble the length of the motorway to be constructed (Fig. 11.10). Shallow tunnels cut through the spurs, alternating with high viaducts; the ground problems are of considerable complexity including the foundations for viaduct piers in weathered rock and sloping ground, the support of tunnels in weathered rock, and the protection of the works from landslides both during and after construction. Heavy rain is likely to cause catastrophic slope failures, as is illustrated by Fig. 11.11. The gully was filled, and subsequently cleared, of debris. In this case the gully was to be crossed by a viaduct which would be endangered by subsequent slips if not

Fig. 11.11. Detail of slope on Rodovia dos Imigrantes motorway near site of landslide.

suitably protected. The tunnel portal shown is a protection against rock falls; it is exposed into free air because of differences between the original topographic survey and that established more precisely during construction.

4.2. Implications of changed conditions encountered during construction

If conditions different from those expected during investigation are encountered in construction then they must be taken into account by modification of design and construction procedure. There are two aspects to this problem: first, that which affects the design, and secondly, that which influences the contractor in his progress and so may give rise to a contractural claim for reimbursement of additional costs. In the simplest case, the depth of excavation to achieve a satisfactory foundation may be greater than was expected. For example, early in the construction of the Roseires Dam (Fig. 11.9) it was established that weathered rock occurred deeper than was originally expected. The geological explanation was that weathered rock existed below flat-lying granite sheets intruded into micaceous metamorphic rocks, and many of the original boreholes had terminated in sound, unweathered granite (Knill and Jones 1965). The engineering solution was simply to excavate the foundations of the buttresses to greater depth, relying on additional boreholes to check the depth of weathering. Where a buttress dam is being built within a relatively steep-sided valley such a situation can be encountered because rocks on the valley sides are weathered; additional excavation can be carried out and the appropriate costs for the excavation and concrete paid to the contractor. However, on such a restricted site there will be interference in contractual programming, for different operations will come out of phase with one another and the contractor may claim that the changed conditions have contributed to such additional costs as have arisen.

During the construction of the Tyne Vehicular Tunnel conditions were encountered which were believed to be different from those which had been expected (Open University 1976). The rock section of the tunnel is driven through Coal Measure shales and sandstones; during the driving of the pilot tunnel, loose ground, high water inflows, and two major collapses took place. Subsequent studies suggested that these conditions were a consequence of past mining activity which had contributed to strains and a consequent deterioration in rock properties. Up till that time it had been tacitly assumed that once subsidence had ceased the consequences of deep coal-mining had been essentially eradicated from the geological history of the rock. Later investigations, for the Liège Metro, have confirmed and extended these observations. Situations which give rise to contractual claims commonly focus on those matters which the contractor is expected to carry out at his own cost and allow for within his other rates, such as support of tunnel roofs, dewatering of excavations, and angle of batter of slopes. Claims can be of considerable importance in that they pose important questions as to the adequacy of investigations and the related interpretation, thereby centring on the heart of the functions of the engineering geologist.

5. Operation

Ideally, if all the problems have been recognized and coped with during construction the difficulties which occur during operation should be minimal. With large schemes, it would be unusual if some additional works were not required during the initial phases of operation. For this reason, it is normal practice to instrument large structures so that imposed loadings can be observed and the responses monitored. Such instrumentation will provide advance warning that some treatment of the foundation or structure might be required. The particular problems which occur include excessive settlement associated with under-provision of foundation support, modification to groundwater flow associated with leakage or instability, and slope failures. Remedial works include support, drainage, grouting, and re-grading.

On a larger scale it is necessary to consider the failure of large structures, such as dams, which can result in major catastrophies. The Malpasset Dam, a highly stressed arch dam 66 m in height, failed in 1959 as a consequence of the over-stressing of the left bank, which was composed of schists and gneisses. The failure was localized by a wedge-shaped mass of rock on which the dam was founded. Other dam failures can be a consequence of groundwater flow through earth fill or foundation materials, resulting in internal erosion, or slips induced by weak foundation materials below embankment dams. Interactions between engineering construction and the natural environment are important, as is clearly illustrated by the consequences of reservoir-induced seismicity and the catas-

Fig. 11.12. Rock falls in Hawkesbury Sandstone marginal to Burragorong Reservoir upstream of Warragamba Dam, New South Wales.

trophic slide of part of the northern side of Mount Toc in northern Italy into the Vaiont Reservoir in 1963. There may also be interaction between mineral extraction and construction; the failure of the Baldwin Hills Reservoir (p. 299) provides one example, and the large-scale rock falls resulting from undermining of cliffs marginal to the Burragorong Reservoir in New South Wales another (Fig. 11.12). Natural hazards provide a further, and final, example of the risks to which a structure, or its foundations, may be exposed to during operation; it would be unwise to assume that the engineering problems of a structure cease with its construction.

References

Anon. (1970). The logging of rock cores for engineering purposes. *Q. J. Eng. Geol.* **3**, 1–24.

Carter, P. G., and Mills, D. A. C. (1976). Engineering geological investigations for the Kielder tunnels. *Q. J. Eng. Geol.* **9**, 125–41.

Edwards, R. J. G. (1972). The engineering geologist in project reconnaissance and feasibility studies. *Q. J. Eng. Geol.* **4**, 283–98.

Knill, J. L., and Jones, K. S. (1965). The recording and interpretation of geological conditions in the foundations of the Roseires, Kariba and Latiyan dams. *Géotechnique.* **15**, 94–124.

The Open University (1976). *Earth science topics and methods: urban geology case study.*

12. Geological hazards

1. Introduction

'Hazard' is one of those useful words which means much in general and little in detail. It describes a result; provides us with a measure of potential disruption; and, being gregarious in application, allows us to view the onset of virtually any event as a hazard to that already in existence. As such, every geological process can be considered as a hazard to the products of former times. Indeed, the whole of geological history could be taken as a record of nature's response to hazards that have come to pass. We could have called them geological hazards, but do not, because we prefer to see every change, no matter how violent or catastrophic, as a natural stage in a larger sequence of events. We live on a restless Earth, yet, as soon as man is affected, these self-same changes become clearly recognized as geological hazards. So, rightly or wrongly, man has become an integral part of any situation that would be generally recognized as constituting a geological hazard. Such a hazard will therefore be taken as any situation which endangers man and results from geological processes.

Two basic sources of hazard exist: that produced by purely natural processes (Fig. 12.1) and that induced as a natural response to man's activity. Of the two sources, it seems likely that the size and frequency of the hazards induced by man will increase at a greater rate than those produced by nature alone, and it is not unreasonable to predict that the incidence of geological hazard will only decrease when a level of geological education succeeds in protecting man from himself.

The fact that man is an integral part of our notion of geological hazard greatly complicates its quantification: man does not provide a simple basis for measurement. For example, a man may live, on average, for 60 years (one scale), but the home he occupies may

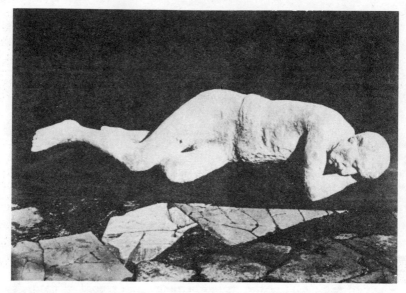

Fig. 12.1. Petrified body, Pompeii (A.D. 79). Geological processes operating at the surface of the earth kill relatively few people; many lose their lives through the collapse of engineering structures. However, this victim probably died from asphyxiation. (By courtesy of Radio Times Hulton Picture Library.)

have an average age of 150 years (another scale), and that home may be part of an economic community that has required 300 years to develop (yet another scale). Regrettably, the only feature common to all men, apart from birth, is death, and fatality is the ultimate means of comparing the effect of one hazard with another. However, this is an abominably crude yardstick, for the injured, the homeless, and the destitute have also to be considered.

Geologists working in this field must therefore be able to assess whether a geological process or reaction will kill, injure, render homeless, or remove the means of survival. This requires the geologist to have a lucid appreciation of the processes involved, coupled with a competent understanding of the geological arena in which he works. The assessment of geological hazards is therefore one of the most skilled tasks that a geologist can undertake.

2. A perspective
Although the nature of a hazard can vary greatly, a prediction of its

possible effect is normally required in the simplest terms: how far? by how much? and for how long? It is therefore sensible to keep any assessment of geological hazard in equally simple terms and, for this purpose an attempt is made here to consider geological hazards in units of length, mass, and time. For a more conventional treatment see Bolt *et al.* (1975). The scale chosen is based on human activity, and three basic divisions have been adopted:

micro: on the scale of an individual and his immediate environs (such as his village or town).

meso: on the scale of groups of micro-units (as in a city and its suburbs, or in a farming community).

mega: on the scale of groups of meso-units (as in a populated region of a country).

Length. Consider the question of distance. Will an earthquake in Alaska be a hazard to a miner in the Appalachians? Will the subsidence affecting Venice reach Treviso and Padua? Will the local disposal of waste pollute existing well supplies? The answers to these questions require a knowledge of physical size and continuity. On a micro-scale we are dealing with joints, bedding, cleavage, and local variations in rock-type and water level. On a meso-scale we have to consider larger features such as individual faults, folds, volcanoes, deltas, major differences in rock-type and stress *in situ*, together with groundwater and other pore contents. On a mega-scale we are dealing with very large geological features such as continental plates, the fold belts and rifts within them, the extensive sequences of igneous and sedimentary rock, and the regional aquifers associated with them. Other geological features will probably have to come to mind, and this should be no surprise, for a comprehensive list would include all geological features of any size. Length is probably the easiest dimension to get into perspective. What of mass?

Mass. There are two reasons why the dimensions of mass are difficult to appreciate. First, all mass on earth is affected by gravity which gives it weight. Secondly, as a material property, it is generally quoted for a single phase; solid, liquid, or gas. So we can determine accurately the mass per unit volume (i.e. density) of fresh water, sea water, gas, crude oil, quartz and the like. The material of near-surface geology is, however, rarely single-phase, because rocks contain pores and fissures which may contain liquid and gas. So in many cases it is necessary to think of bulk mass per unit volume (i.e. bulk density). For example, a recently deposited sediment, compacting beneath its superincumbent load, increases its bulk density as the sedimentary grains pack more closely together. Now many natural processes are generated by changes in either the mass of, or the charge on, the matter involved. Two great examples in geology are convection in the mantle and weathering at ground level: the consequences of these processes alone account for a large proportion of all geological studies, yet they both stem fundamentally from changes in mass. So problems that largely involve mass cover a very wide spectrum indeed. For example, will methane in a coal-mine tend to rise or fall in a shaft? ... a simple problem involving the relative mass of gases, methane and air. Will sea water invade a coastal aquifer if fresh water is gently pumped from the ground? ... a slightly more complicated question, involving the relative masses of sea and fresh water. Will natural valley erosion endanger the stability of valley slopes? ... a difficult problem, partially involving the relief of gravity-induced stresses (a function of mass) and weathering. Will an increase in the temperature at the margin of a continental plate increase the incidence of vulcanism? ... a complex problem involving phase relationships at different depths.

Now on a micro-scale, we are dealing with local problems which often result from local causes, and can be modified usually by a few men: a farmer can treat the ground so that the disadvantageous effects of weathering can be reduced to a minimum, and a coastal water supply can be managed so as to prevent pollution by sea water. A meso-scale problem can be recognized when it is no longer possible for a few men to control events, and a significant communal input is required to safeguard man from some hostile situation: one example is the protection of small communities from volcanic eruptions. This, however, is approaching the scale of a mega-problem where man is not in control and the natural processes, which generate the hazard, are regional rather than local. In short, the change in material characters brought about by these mega-processes is usually so great that man simply cannot cope with it for any period of time.

Time. Man lives on a crust that moves more quickly in some parts than in others, where such movements can last longer in some places than in others, and occur either regularly or irregularly with time. Hence, plans to combat geological hazard must take into account the rate of change to be expected, the duration of this change, and its frequency. These are three forms in which time can be considered. Of them, rate of change is the most important to man, for this largely determines the success of evacuation. So micro-time problems are taken as those involving a rate of change that lies comfortably within the range of man's mobility; creep is an example. Meso-problems of time occur when the rate of change approaches the limits of man's mobility, in that a man could probably save his life but not his property; many landslides produce this problem. Mega-problems involve a rate of change that will cause either death or serious injury; violent earthquakes are a common example.

TABLE 12.1

A synopsis of criteria that can be used for defining geological hazards

Scale	Dimension	Character of hazard
Micro	Length	Local variations of small extent and continuity.
	Mass	Limited disturbances that can be dealt with by a few men.
	Time	Rates of change within the range of man's mobility.
Meso	Length	Variations which influence many people's habitat.
	Mass	Disturbances which can only be controlled by a communal effort.
	Time	Rates of change that approach the limits of man's mobility.
Mega	Length	Variations which affect complete regions.
	Mass	Disturbances which are beyond the control of man.
	Time	Rates of change which normally result in injury and death.

Two points emerge from this perspective. The first is that statistics have not been used to define whether a problem is large or small; see Table 12.1 and, for comparison, Scheidegger (1975). It is largely one's personal relationship to the problem that will determine whether this view is correct. The second point is that the basic divisions in Table 12.1 merely reflect a common approach to alleviating the effects of hazards. For example, a developed society would recognize micro- and meso-problems as largely a matter for insurance, and mega-problems as 'acts of God'. A developing society would see micro-problems as mainly a question of experience and meso-mega-problems as a matter of luck. In other words, man can live with problems on a micro-scale. Consequently, it seems best to direct our limited efforts in the field of geological hazards to the reduction of mega-problems to a more acceptable meso-scale, and meso-problems to a more manageable micro-scale. Such a reduction will become increasingly possible as the causes and behaviour of these hazards become more fully understood. The first problem here is to identify hazard-producing processes.

3. Identification of natural hazards

If it is accepted that natural geological hazards stem from geological processes which adversely involve man, it follows that the identification of naturally produced hazards will rely basically on an understanding of the geological processes themselves. Significant advances have already been made in this field, but their value is currently enhanced by the fact that geologists can now obtain an absolute value for age and hence time. Formerly geological time was excluded from the discussion of time but that did not mean it was unimportant, because its product, geological history, provides us with a unique source of reference for events which have been possible in the past; obviously, field geology will continue to be fundamental to the advances made in relieving geological hazards. The greatest advances come, however, when field and laboratory combine to produce a definition of equilibrium, for the increased sophistication of our instrumentation and analysis has revealed that geological processes are remarkably sensitive to changing conditions. It is during the periods of change from one equilibrium condition to another, that most naturally produced hazards arise. Two examples will illustrate this; both involve near-surface processes, since these are more clearly understood than those at depth, which cannot be observed.

Quick-clays. In Scandinavia and Canada there are extensive deposits of marine clay that accumulated during and after the Pleistocene. Being marine, the clay particles were deposited in a flocculated condition and, despite their high porosity, developed a moderately large shear strength from the bonds formed by the edge-to-face contact of the clay plates. These deposits were not subjected to later burial, but were raised above sea level by isostatic uplift of the continents on which they lay. From that time on, they have been flushed by percolating meteoric water, which has progressively removed salts from the pores of the clay and reduced the concentration of electrolyte which surrounds the clay particles and their points of contact. This has reduced the net attraction between the particles and placed an ever-increasing load on bonds that developed during the years of confining stress. When these are broken, as can occur with some disturbance of the ground, the clay may lose all its strength and become a soil–water slurry. This final loss of strength can happen very quickly. Here, man is involved in a natural transition between two equilibrium conditions (from flocculated structure to dispersed) that is taking thousands of years to complete.

Flood run-off. Excessive run-off can occur when the rate of precipitation exceeds the rate of infiltration, and this incidence is likely to increase in temperate climates, when precipitation depletes the storage capacity of the ground. Most catchments support a soil profile and this has characteristics similar to a sponge. The pull of gravity will always induce the drainage of water within it, until the capillary force within the pores balances the gravitational force acting on the water; equilibrium is then established, and the soil is described as being at its field capacity. At this point there is no soil moisture deficit. Now, vegetation rooted in the soil draws on soil–water and is capable of exerting a greater 'suction' than gravity, so that dewatering can continue beyond the point at which soil is at its field capacity. This produces a soil-moisture deficit. Thus vegetation generates storage space in the soil when transpiration exceeds precipitation. This storage is progressively eliminated, when precipitation exceeds transpiration, as each rainfall can reduce a little of the deficit that has accumulated during drier seasons. Soil-moisture deficits can therefore cushion the response of a catchment to heavy rain, and it is known that the risk of flooding increases sharply once the soil has been restored to its field capacity. Here we have an ever-changing sequence of events, and hazards arising when geology can no longer provide the equilibrium conditions that permit adequate rates of infiltration.

Both these examples illustrate mega-problems that can be and, where necessary have been, reduced to a meso-scale by an appreciation of the processes at work. Engineers can cope with quick-clays (Bjerrum 1967) and the likelihood of flooding occurring in a catchment unexpectedly can be reduced (Beran and Sutcliffe 1972). Understanding the equilibrium of natural systems is the key to success and the same key will reduce the hazards induced by nature's response to man.

4. Identification of induced hazards

Subsidence resulting from the extraction of oil, gas, and water is just one example of how man can superimpose a regime on that already existing and upset the progress of geological processes that are working towards some natural condition of equilibrium. This interference has, for a long period, been on a micro-scale but is now conducted on a meso-scale and, at these levels, it can seriously disturb geological processes that are close to limiting equilibrium, and so precipitate a response which might not have occurred as rapidly, or at all, under natural conditions. A simple example of this is found in landslides. The shear strength of a surface that has little or no cohesion across it is given by $S = (N - u) \tan \phi$, where N is the total pressure acting normally to the surface, u the water pressure on the surface (acting in all directions), and $\tan \phi$ the coefficient of friction. If the surface is dipping out of the slope at an angle that brings the forces acting on it to limiting equilibrium, then a small increase in (u) will promote failure: i.e. any interference that could produce a small water-level rise in the slope could start a landslide (Hoek and Bray 1974). Thus a knowledge of both the geological processes and the nature of man's interference is required if the geological hazards which can be induced by man are to be successfully identified. Unfortunately, man rarely knows in detail what he is doing to the ground, so that when a hazardous situation occurs it is difficult to analyse its origins. The identification and avoidance of induced hazards will necessitate greater sums of money being spent on monitoring ground performance before, during and after man's interference (see, e.g., MacLean 1966). However, even neglecting details, it is possible to identify three engineering situations which should alert both the geologist and the engineer to the dangers of inducing a hazardous ground response: they are schemes which involve (i) an increase in crustal load, (ii) a decrease in crustal load, and (iii) a change in hydrology.

Increase in crustal load

The vertical load, acting on an element at depth, can be expressed as $L = (N - u)$, where N and u have the same definition as given in the previous paragraph. All structures increase the load on the crust, but it is the very large structures and widespread processes, approaching a meso-scale, which have the most dangerous influence (Fig. 12.2.). This is currently occurring on a large scale in the extraction of oil, water, and gas, and the impounding of large surface reservoirs.

The extraction of fluid from depth reduces pore pressure (u), and so increases effective stress ($N - u$) and encourages gravitational

Fig. 12.2. Underground collapse in Coal Measures. Heat, pressure, water, and gas are the immediate geological hazards below ground. Here ground pressures have crushed a supported roadway; methane was released but has dispersed. (Photograph by courtesy of the National Coal Board.)

compaction in materials which have interparticle bonds too weak to support the resulting increase in load. In general, the response of a sedimentary material to a reduction in pore pressure can be expected to reflect its position within the overall sequence of sedimentation, burial, and exhumation. During deposition the fabric of a new sediment is open and delicate, readily responding to any changes in effective stress, regardless of cause. As the process of sedimentation continues, the deposit gains some strength, but nevertheless, is vulnerable to changes in pore pressure. The subsidence in Tokyo is an example. The city is partially founded on poorly consolidated Quaternary deposits, from which water is extracted. Subsidence began about 1920 and has continued, so that the maximum settlement is now approximately 14 ft, the lowest ground being 7 ft below sea level. This subsidence affects an area of almost 30 square miles, which is inhabited by over 2 million people.

During burial, pore pressures are being both generated and dissipated; the two processes may balance each other but excessive formation pressures found at depth illustrate that this need not occur. A reduction of these pressures can induce a marked response. Yerkes and Castle (1969) report that differential subsidence, centripetally directed horizontal displacements, and faulting can all occur with the extraction of oil and gas from depths in excess of 3000 ft. Differential subsidence is the most common and widespread effect which centres over, but extends well beyond, the producing area. Horizontal movements usually accompany subsidence. Faulting is not so common but nevertheless frequently occurs, often suddenly. Most faulting is high-angle, normal, and peripheral to the subsidence bowl, with down-throw occurring on the production side of the fault. Very occasionally, low-angle reverse faults occur, centrally positioned in the basin of subsidence. It is interesting to note that the strain pattern developed over oil and gas fields are similar to those that have been measured over a Texas salt dome. Here, all three surface effects were observed during extraction of sulphur by the Frasch process.

The process of exhumation destroys some of the features produced by burial; structures expand as the load upon them decreases, and pores and fissures enlarge, often to be filled with liquid and gas. Subsequent reductions in the pressure of these pore contents will enhance compaction, though not to the extent found in the situations mentioned above.

Now even this extremely simple exposition of geological environment helps to clarify our knowledge of the hazards that can be induced. Significant compaction or settlement can be expected in newly formed sediments but, by reason of their plasticity, their rate of change is likely to be gradual and fairly uniform. This plasticity can be lost in deposits that have been buried, so that although the resulting compaction may not be so great, it is likely to occur less uniformly in both space and time. Poland (1971) reviews many instances of subsidence throughout the world which reflect these trends. Permeability will develop markedly anisotropic characters and definite geological structures, such as faults and folds, will be forming. All this will influence the response of the ground. The importance of these structures remains until the stress relief on exhumation reduces their effectiveness, so that variations in such characteristics as transmissivity and storage reflect variations in topography rather than structure and stratigraphy.

The impounding of large surface reservoirs increase the normal load N acting on the crust by an amount equal to the height of water impounded. Since many dams impound to depths of less than 200 m (100 m of water produces a pressure of 10 bars), it seems unlikely that such loading would significantly affect geological equilibrium. However, a growing volume of evidence suggests that reservoirs can promote ground adjustments which result in earthquakes, for approximately one in a hundred man-made lakes experience an increase in local seismic activity when filled (Howells 1973). The examples given here come from Monteynard and Koyna.

Impounding of the Monteynard dam in the French Alps began in April 1962 and the reservoir was filled in April 1963. That month a shock of magnitude 5·0 occurred, with its epicentre at the dam. A total of 15 shocks were recorded in 1963, 10 in 1964, 23 in 1966 (one of magnitude 4·3), and 16 in 1967. The strongest shocks occurred when the reservoir was at its maximum height, with a depth of water just short of 130 m.

The Koyna dam is founded on the rocks of the Deccan Plateau, which is one of the least seismic areas in the world. Impounding commenced in 1962 and shocks occurred a few months later; the epicentres were calculated as being near the dam and under the reservoir. Filling was completed by 1974, but the shocks continued, with that of December 1967 (magnitude 6·4) killing 177 people and wounding 2300. Numerous after-shocks followed, that of 1968 having a magnitude of 5·4. The maximum depth of water in the reservoir was about 100 m.

Rothe (1973) records many examples of a correspondence between impounding and seismic activity. In many respects each case is unique; but when considered collectively they reveal some common characteristics. For example, seismic activity became pronounced once the depth of water exceeded 100 m (lowering the water level in reservoirs has halted seismicity and raising the level has restarted it). Often this activity has produced its most violent shocks after a period of numerous fore-shocks. It is now generally accepted that this seismicity resulted from disturbing a geological situation that was close to limiting equilibrium. Future identification and prediction of this hazard virtually opens a new field in reservoir investigation.

Decrease in crustal load

Mining is probably the most common source of artificial large-scale crustal unloading. But it is not the only source, for since the 1940s there has been an increasing use of deep fluid injection; this raises pore pressures u at depth and so lowers effective stress $(N-u)$. Deep fluid injection was begun on a large scale in the oil industry as a means of improving the yields from old and difficult fields. Since then it has also become a method increasingly used for disposing of unwanted fluid. The effects of deep fluid injection and mining can reach a meso-scale.

Opencast mining invariably prompts some response from the ground, usually in the form of rebound or floor heave. In some cases this can be dramatic; Coates (1964) records the sudden formation of a dome 8 ft high, 50 ft long, and 90 ft wide in the floor of a limestone quarry. However, opencast operations are usually fairly restricted; of larger scale are the effects of underground mining, especially subsidence. This can generate many hazards at ground level by damaging dwellings, sewers, and piped supplies. But there are greater hazards still, often unseen and generally associated with water. This is discussed under the heading 'Changes in Hydrology' and it is sufficient to note here that the zones of tension and compression that accompany subsidence can act as migrating groundwater boundaries, and that subsidence itself can increase the areas susceptible to inundation by surface flooding.

Subsidence is a good example of a ground response that is understood well enough for engineers to forecast the magnitude of future ground disturbance. Hence structures can be designed to withstand later movement of their foundations (National Coal Board 1966); but this has only become possible because thousands of measurements had been made previously of ground movement. If this is the level of information required to predict, in usable terms, hazards which can be induced it is clear that we shall be unable to 'live' with many of the hazards which may have characteristics as yet largely unmonitored.

TABLE 12.2

Summary of events associated with the ground failure at Baldwin Hills

Crack No. (see Fig. 12.3)	Occurred or first observed	Length (ft)	Year	Volume injected near Reservoir fault (barrels × 10^6)
1	May* 1957	2600	1957	0·8
2	Jan. 1958	1900	1958	1·1
3	Mar. 1958	600	1959	1·2
4	Before 1961	400	1960	1·8
5	Feb–Mar. 1963	400	1961	4·0
6	1962–3?	800	1962	3·0
7	Feb. 1963	400	1963	4·4
8	Aug.–Dec. 1963	2000	1964	4·6
9	Aug.–Dec. 1963	800		
8	1968 ⎫ recurrent			
9	1968 ⎭ movement			
10	1968	1400		

* First large-scale injection well commenced in May 1957.

However, subsidence is not the only hazard to result from unloading below ground; stress relief can cover the whole scale of time from mega to micro; for although subsidence may be slow, rock-bursts are rapid, lethally so for miners (Obert and Duval 1967). The other form of crustal unloading (fluid injection), although new compared with mining, has already been associated with some alarming situations. Two are described: the failure of Baldwin Hills reservoir and the Arsenal Well at Denver.

Baldwin Hills is a suburb 8 miles from Los Angeles. The hills themselves are formed by the Inglewood anticline which is faulted and oil-bearing. The oil was discovered in 1924 and rapidly developed, with fluid pressures in the producing zones dropping from approximately 507 lb/in² to 50 lb/in². This drop was not uniformly transmitted throughout the field, for the faults acted as hydraulic boundaries, which enabled some compartments to retain higher pressures than their neighbours. Subsidence accompanied the reduction in pore pressure and was first noted in 1943. In 1951, a surface reservoir was constructed in the hills, and operated satisfactorily even though subsidence continued. However, by 1954 most of this subsidence had occurred, and in May 1957 a full scale programme of water flooding (injection of water into the oil-bearing horizons) was started to improve secondary recovery. In that same month, cracks began to appear in the vicinity of the reservoir; Table 12.2 and Fig. 12.3 illustrate the history from that date. On 14 December 1963 a 7-inch upward movement on the Reservoir fault permitted water to escape through the embankment of the reservoir. 250 million gallons of water flooded into Baldwin Hills,

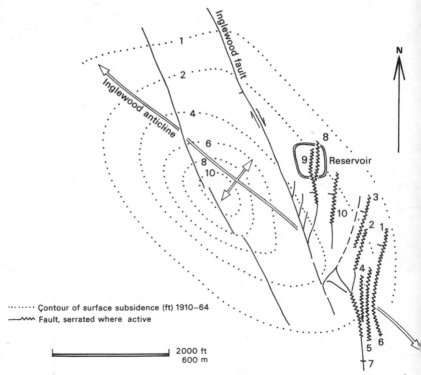

Fig. 12.3. Location of Baldwin Hills Reservoir relative to Inglewood anticline. (Based on Hamilton and Meehan 1971.)

damaging and destroying 277 houses. Evacuation had been possible and only 5 lives were lost. Six months previously there had been 9 injection wells operating near the Reservoir fault, and these had experienced uncontrollable losses of fluid and shearing at depth. In addition, injected brine had been seeping up to ground level along the trace of this fault. Hamilton and Mechan (1971) believe that failure was basically caused by fluid injection, whereas Castle, Yerkes, and Yound (1973) present evidence to show that fluid injection merely accelerated a failure that was *already* in progress, a failure that stemmed originally from the withdrawal of fluid and the lowering of pore pressures.

The Denver well was used for disposing of fluid waste from the Rocky Mountain Arsenal. The well was 12000 ft deep and penetrated fractured crystalline Pre-Cambrian rocks at 11950 ft. Injection occurred through the lower 50 ft, was intermittent and never exceeded a pressure of 1000 lb/in^2 and a rate of 300 gallons/min. The chemistry of the waste was not known in detail, but it was largely a solution of NaCl (13000 ppm), having an average pH of 8·5. Injection commenced in 1962 and was accompanied by a spate of small earth tremors; none had ever been recorded in the same area before. Injection continued, each period of operation being followed by a period of seismic activity. Public concern finally stopped the operation of the well in 1966, but the quakes continued with three large events of magnitude of 4·8, 5·1, and 5·3 occurring in 1967. By that time it had become apparent that the epicentres associated with these disturbances had migrated 4 to 5 miles from the well. Furthermore, they were located at a depth of 2 to 3 miles and were thought to be originating from movement along a shear-zone. Although it seemed certain that injection had triggered off these events by reducing effective stress at depth, it was not certain whether this alone explained the continued activity long after the well had ceased to operate. Various explanations have been put forward (van Poolen and Hoover 1970); they can be summarized as follows: (1) The coefficient of friction of the sliding surfaces was decreasing from its maximum to its minimum value, so that movement could continue with ever-decreasing pore pressures. (2) The difference in temperature between the waste and the ground at 11500 ft would generate thermal stresses 10 times greater than those produced by the pressure of injection; this, however, is unlikely to cause movements 4 to 5 miles from the well. (3) The chemistry of the waste weakened the ground. It was calculated that 210 tons of reservoir rock was dissolved, and that Na–K exchange involved 320 tons of K-feldspar. Not surprisingly, no single mechanism is thought to have caused the earthquakes.

Changes in hydrology

A number of geological factors can be changed on site to suit the requirements of an engineer. For example, the strength of the ground can be improved and the stresses within it can be re-aligned. But this work is expensive and local, the benefits accruing from it being restricted to areas immediately surrounding the points of treatment. The treatment of groundwater is different, because water flows and, as such, is the one geological factor that man can usually change extensively for comparatively little cost; large volumes of ground can therefore be affected. Consequently, changes in hydrology can promote geological hazards which soon reach a meso-scale. There are three areas where this is commonly happening, namely underground mining, basin management, and land use.

Underground mining can promote a number of changes in the natural hydrology of an area; two will be considered. As mentioned above, ground strain associated with subsidence, can result in migratory groundwater boundaries, which can severely alter the natural pattern of subsurface flow. An example of this has been recorded by Mather, Gray, and Jenkins (1969) from the Aberfan area of the South Wales coalfield. Here tracer investigations revealed that small catchments were linked hydrologically by a corridor of tensile strain which passed through them, so that groundwater no longer remained within the general confines of the surface-water catchments, but transgressed their boundaries; the corridor acted as an underground drain. Furthermore, the tensile strain had increased the permeability of the ground within this corridor by many orders of magnitude. Such changes need not be hazardous, but they are unseen and unheard and, as such, their presence can pass unnoticed. This was the situation at Aberfan when a tip of colliery waste, placed on the valley slopes above the village, failed and flowed, with great speed, down the slope and into the village itself, killing 144 people. The failure of the tip stripped a covering of fairly impermeable clay from the hillside and this permitted the release of water from a permeable corridor of tensile

302

strain which was located beneath the tips. This water, which mixed with tip-material to produce a mud-flow, also entered the village, making an already catastrophic situation even worse.

A second source of hazard comes from the dewatering that is often associated with mining. This can cause settlement at ground level, much of which comes from the consolidation of weak material. Occasionally, as in limestones, the settlement may be due to collapse within sinkholes; such is the case in South Africa, where the gold-mines of the West Rand operate beneath 4000 ft of Pre-Cambrian limestone. Dewatering of the limestones has accelerated the collapse of existing caverns and has claimed the lives of a number of people living on or near the mines; much property has also been damaged. Large areas of land on the limestone so affected have been declared uninhabitable because of the sinkhole risk. This illustrates an aspect of geological hazards that has not yet been mentioned, namely the alarm they cause. When the areas affected by subsidence on the West Rand are assessed, they come to a very small proportion of the area proclaimed uninhabitable. In fact, the chances of meeting injury or death in a sinkhole collapse are small but, naturally, the people involved are not particularly interested in statistics; facts are more to their liking, particularly geological evidence of firm ground. The problems produced by groundwater on the West Rand illustrate so many geological principles that engineers and geologists associated with mining will find the reports of Cartwright (1969) and Cousens and Garrett (1969) of particular interest.

The second potential area of hazard, basin management, entails controlling the movement of water in a catchment. Virtually nothing can be done to control rainfall but the remainder of the hydrological cycle (run-off and infiltration) can be manipulated to some extent. One management facility that is currently becoming necessary is the river barrage. This can protect a valley from tidal flooding, impound fresh water, and even generate electricity (see e.g. Cotillon 1974). However, barrages usually produce an afflux, which in turn raises the level of water in surrounding ground; this can affect the stability of foundations and the discharge of water to underground services. Gray and Foster (1972) describe the extent of some of these features in that part of London which may be affected by an increase in water level following the closure of the Thames barrage.

Geological hazards

Land use (or abuse) can result in geological reactions which seriously affect infiltration and promote soil erosion. Brown (1972) describes the effect bush fires have had on two neighbouring catchments in New South Wales; the total catchment area was about 280 square miles. The shapes of the flood hydrographs were markedly changed after the fire, and the highest flood in 19 years of record occurred within 10 months of the fire itself, after a period of relatively *low* rainfall. Furthermore, the run-off from the catchments was increased significantly for 4 years after the fire, as was the sediment load of the streams. These catchments took 4 to 5 years to recover their original hydrological characteristics, but many never recover and permanent damage can be done to the livelihood of regions so affected. Here again, hydrology can rapidly produce a hazard that reaches a meso-scale. Young (1973) discusses the problem of using soil surveys in land development, and analysed the character of many soil surveys that are commissioned (but see also Wermund *et al.* 1974). Young's conclusions are worth noting, for they reveal that often the effort entailed in a survey has been misdirected. Would a similar conclusion emerge from a study of the geological investigations that were undertaken for contracts which later induced geological hazards?

5. Forecasting

Identification of geological hazards, of either natural or induced origin, is of limited use to man, as some forecast of future behaviour is also required. This takes two basic forms: will it ever happen? (Question 1), and, if so, how quickly? (Question 2). Both questions deal with rate of change, which is the parameter already chosen for scaling problems involving time. For example, Question 1 means, 'At what rate will the identified hazard develop to a critical condition?', and Question 2 means, 'At what rate will the critical condition change our existing norms?'. These questions help to define the direction in which the effort of forecasting can best be used, for time, or rate of change, is an important criterion in deciding both the object of a forecast and the level to which it should be pursued. There is limited benefit in forecasting the nature of events in movements that occur so quickly that their rates of change normally result in injury and death; what does it benefit a miner to know the sequence of events during an explosion if the explosion takes only a fraction of a second? Here it would be better to forecast the amount of time it takes for the hazard of explosion to develop to a critical condition. In other words, the effort involved in forecasting geological hazards which involve mega-problems of time (Table 12.1) would probably be better spent studying when it will occur (Question 1) rather than how quickly it will happen (Question 2). Both would be studied for hazards that produce a slower ground response, although in many of these cases it might be preferable to devote the greatest effort to Question 2 (see e.g. Foster and Crease 1974).

Forecasting is the acid test of our understanding and attracts few devotees, yet we are used to working with forecasts almost every day, the most common for most of us being that for weather. Weather forecasts are good when a settled, monitored system is in operation, are very good when such a system is repeatable, but fail when conditions become unusual. But 'unusual' means 'not often experienced', and so our understanding of the 'unusual' changes as a record of events increases. Now geologists have an immensely greater record to consult than meteorologists: the geological column is their source of reference, and this has been used, for geologists can decipher the order of events which occurred millions of years ago. Indeed, by far the greater part of their training takes them back in time to make predictions (i.e. forecasts) of the long-distant past, by using the principle of uniformitarianism: 'the present is the key to the past'. But can geologists go forward with the same measure of success? Is the past the key to the future? If it is, then we have the greatest storehouse of all the earth sciences upon which to base our forecasts. Examples of this approach are occasionally seen; for example, Walker (1974) has noted that civilized man has not yet experienced a large ignimbrite erruption, such as can be expected to occur somewhere in the world, on average, once every 10^3 years; because of this, the nature of the premonitory signs which will enable such a cataclysmic event to be predicted can, at the present, only be surmised. Under these circumstances the past is only *one* key to the future. Finding the other keys will therefore require the application of techniques that go beyond the bounds of our present approach.

6. Future trends

Numerous trends can be visualized, particularly in the collection and storage of data, but the most important trend will probably require the substitution of geological time, for one of the greatest problems in appreciating the rates of geological processes is the time these rates take to change. Most geological transitions operate for the greater part of their duration at slow rates of change and it is the occasional spurt that often causes the hazard. How then, can such a system be studied? Present facilities and training ensure that the geological record can be studied in reasonable detail and used as a source of hypotheses for Earth science. But to study and test these hypotheses it will become increasingly necessary to reproduce the relevant geological systems by models; once a model is made, its rate of change can be controlled by an operator. This replacement of geological time by operator time will make it possible to apply a wide range of basic geological studies to improving the quality of life.

The foundations for such an advance have been laid already by management and industry in their study of cybernetics (man–machine interactions) and systems analysis. Perhaps these words should not be used for geological studies, since this application is far removed from their original definition. A growing number of earth scientists are, however, finding that the logic used in cybernetics and systems analysis is pertinent to the solution of current geological problems (Griffiths 1970; Buttner 1972). For example, to forecast a particular geological hazard it may be necessary to study a number of geological processes (such as the distribution of stress in rock, and rock breakage) and their reaction to a certain type of man-made interference. It may then be necessary to link these reactions together into a system so as to study the outcome of this system, i.e. to see what happens when the various geological reactions are linked together by a numerical law to produce a system whose progress is governed by the reactions generated within the system. The slope failure at Vajont reservoir, which killed more than 1500 people, would be a typical example of this application (Jaeger 1969).

The pressures for and from such trends could influence geological studies in three basic ways.

(1) Greater effort would have to be devoted to the study of rates of change in geology. This in turn would mean re-analysing parts of the geological record and closely observing geological processes that are operating at the present, e.g. conti-

nental drift, seismic and geothermal activity; see for example Saemundsson (1974), the Royal Society (1973), and Sherard, Cluff, and Allen (1974).

(2) It would become necessary to understand basic geological processes sufficiently well to enable them to be described either by computational or analogue techniques; see for example Collinge (1972) and Domenco (1972). This will require the back analysis of the geological processes observed.

(3) The sequence of events involved in large-scale adjustments of geological conditions, that is, those that are large by man's scale, would have to be determined, so that the order in which geological processes occur could be known accurately; see for example Lomnitz (1974).

All these trends can be seen in work such as that described by UNESCO (1971), Handin and Raleigh (1973), Rikitake (1974), and Press (1975).

Future trends leading to fundamental advances therefore appear to lie in the fields of basic geological investigation and numerical analysis. But what about manpower? There is a great need for geologists to become more involved in the decisions that are made in planning and development, and this need demands geologists who appreciate the analytical aspect of their subject. In addition, there is a complementary need for engineers who appreciate the geological content of their designs. The extent to which these requirements are satisfied will reflect the level of geological education that has been achieved, for there is unlikely to be sustained progress in the reduction of geological hazards if this level of education is allowed to stagnate. Indeed, the problems that are now being encountered indicate that geological hazards are likely to increase unless the geological education of engineers is improved.

References

BERAN, M., and SUTCLIFFE, J. V. (1972). An index of flood-producing rainfall based on rainfall and soil moisture deficit. *J. Hydrol.* **17**, 229–36.
BJERRUM, L. (1967). Engineering geology of Norwegian normally-consolidated marine clays as related to settlements of buildings. *Géotechnique* **17**, 81–118.

Geological hazards

BOLT, B. A., *et al.* (1975). *Geological hazards*. Springer Verlag.

BROWN, J. (1972). Hydrologic effects of a bushfire in a catchment in south-eastern New South Wales. *J. Hydrol.* **15**, 77–96.

BUTTNER, P. (1972). Systems analysis and model building. In Symposium on Quantitative geology. *Spec. Pap. geol. Soc. Am.* **146**, 69–86.

CARTWRIGHT, A. (1969). *West Driefontein—ordeal by water* J. G. Ince & Son (Pty) Ltd., Gold Fields of South Africa Ltd.

CASTLE, R., YERKES T., and YOUD, T. (1973). Ground rupture in the Baldwin Hills—an alternative explanation. *Bull. Assoc. Engineering Geologists.* **10**, 21–49.

COATES, D. R. (1964). Some cases of residual stress effects in engineering work. In *State of stress in the earth's crust.* pp. 679–88, Elsevier, New York.

COLLINGE, V. (1972). Mathematical models of water resource systems. In *Symp. on advanced techniques in river basin management. Birmingham 1972.* Institution of Water Engineers.

COTILLON, J. (1974). La Rance. Six years of operating a tidal power plant in France. *Wat. Pwr.* **26**, 314–22.

COUSENS, R., and GARRET, W. (1969). The flooding at the West Driefontein mine. *Proc. 9th Commonw. Min. Metall. Congr.*

DOMENCO, P. (1972). *Concepts and models in groundwater hydrology.* McGraw Hill, New York.

FOSTER, S. O., and CREASE, R. I. (1974). Nitrate pollution of chalk groundwater in E. Yorkshire:- a hydrogeological appraisal. *J. Instn Wat. Engrs.* **28**, 178–94.

GRAY, D. A., and FOSTER, S. S. (1972). Urban influences upon ground-water conditions in Thames flood plain deposits of Central London. *Phil. Trans. R. Soc. A* **272**, 245–57.

GRIFFITHS, J. (1970). Current trends in geomathematics. *Earth Sci. Rev.* **6**, 121–40.

HAMILTON, D., and MEEHAN, R. (1971). Ground rupture in Baldwin Hills. *Science, N.Y.* **172**, 333–44.

HANDIN, J. V., and RALEIGH, C. (1973). Man-made earthquakes and earthquake control. Paper T2–D. *Proc. Percolation through fissured rock. symp.* Int. Soc. Rock Mechanics and Int. Assoc. Engineering Geology. Stuttgart.

HOEK, E., and BRAY, J. W. (1974). *Rock slope engineering.* Institution of Mining & Metallurgy, London.

HOWELLS, D. E. (1973). Man-made earthquakes; the reservoir designers problem. Paper T2–E. *Proc. Percolation through fissured rock. symp.* Int. Soc. Rock Mechanics and Int. Assoc. Engineering Geology. Stuttgart.

JAEGER, C. (1969). The stability of partly immersed fissured rock masses and the Vajont rock slide. *Civ. Engng publ. Wks Rev.* 1204–207.

LOMNITZ, C. (1974). *Developments in geotectonics. Vol. 5: Global tectonics and earthquake risk.* Elsevier, Amsterdam.

MACLEAN, R. (1966). Foul air in wells and boreholes in the London area. *Proc. Soc. Wat. Treatm. Exam.* **15**, 271–83.

MATHER, J. D., GRAY, D. A., and JENKINS, D. G. (1969). The use of tracers to investigate the relationship between mining subsidence and ground-water occurrence at Aberfan. *J. Hydrol.* **9**, 136–54.

National Coal Board. (1966). *Subsidence engineers handbook.* N.C.B. Production Department.

OBERT, L., and DUVALL, W. E. (1967). Rock-bursts, bumps and gas-outbursts. In *Rock mechanics and the design of structures in rock*, Chapter 19. John Wiley, New York.

POLAND, J. (1971). Subsidence and its control, in Underground waste management and environmental implications. *Mem. Am. Ass. Petrol. Geol.* **18**, 50–71.

POOLEN, H. VAN, and HOOVER, O. (1970). Waste disposal and earthquakes at Rocky Mountain Arsenal, Denver, Colorado. *J. Petrol. Technol.* **22**, 983–93.

PRESS, F. (1975). Earthquake prediction. *Scientific American* **233** (5), 14–23.

RIKITAKE, E. (1974). Focal processes and the prediction of earthquakes. *Tectonophysics*, special issue. **23**, No. 3.

ROTHE, J. (1973). Summary geophysical report. In *Man-made lakes, their problems and environmental effects* (ed. D. Ackerman *et al.*), American Geophysical Union. Geophysical Monograph 17.

Royal Society. (1973). Measurement and interpretation of changes of strain in the earth. *Phil. Trans. R. Soc.* Series A, **274**, No. 1239.

SAEMUNDSSON, K. (1974). Evolution of the axial rifting zone in Northern Iceland and the Tjornes fracture zone. *Bull. geol. Soc. Am.* **85**, 495–504.

SCHEIDEGGER, A. E. (1975). *Physical aspects of natural catastrophes.* Elsevier, Amsterdam.

SHERARD, J., CLUFF, L., and ALLEN, C. (1974). Potentially active faults in dam foundations. *Géotechnique* **24**, 367–428.

UNESCO. (1971). *The surveillance and prediction of volcanic activity.* UNESCO, Paris.

WALKER, G. P. L. (1974). The prediction of volcanic eruptions. *Proc. Geol. Soc.* In *J. geol. Soc.* **130**, 401.

WERMUND, E. *et al.* (1974). Test of environmental geologic mapping. Southern Edwards Plateau, South-West Texas. *Bull. geol. Soc. Am.* **85**, 423–32.

YERKES, R., and CASTLE, R. (1969). Surface deformation associated with oil and gas field operations in the United States. In *Land subsidence* **1**, 55–66. No. 88. Int. Assoc. Sci. Hydrology.

YOUNG, A. (1973). Soil survey procedures in land development planning. *Geogrl. J.* **139**, 53–64.

13. Geology in conservation

1. Conservation

Conservation is concerned both with safeguarding natural phenomena of outstanding scientific importance and the preservation and
improvement of the quality of the environment as a whole. Its field
of potential activity coincides geographically with that of industrial
geology and consequently many contacts arise between the practitioners of the two disciplines. On occasion these contacts are to
the benefit of both, as happens when mineral extraction creates new
habitats or where new exposures of geological merit are opened by
quarries or cuttings. In other cases, conflicts arise, as, for example,
when mineral development threatens to destroy localities of great
conservation value or to pollute the environment at large (Smith
1975). There thus can be no constant relation, whether alliance or
opposition, between the industrial geologist and the conserver of
nature; it is necessary for both to assess each individual case and to
study and respect the viewpoint of the other. This is perhaps more
natural than it might seem at first sight, for industrial geologists
appreciate the benefits of a pleasant and interesting environment as
much as conservers enjoy using the mineral resources of the Earth.

In Britain, as in most developed countries, the results of past
human activity in both fields are clearly marked. The consequences
of the long-continued industrial exploitation of a wide variety of
natural resources are widely distributed and take many forms,
ranging from the stark dereliction of the Swansea Valley to the
holiday attractions based on the disused medieval peat workings
which are now the Norfolk Broads. Conservation also has a long
ancestry, as almost all the environment, whatever its quality, has
been affected to a greater or lesser extent by man. Many of the
most interesting localities for the conserver thus share one characteristic with areas of industrial dereliction: both are man-made.

Conservation seeks the wisest use of natural resources, interpreting the term to embrace not only those, such as mineral deposits, which can be physically quantified but also others, less tangible, which determine the quality of the environment. It differs in its aims and methods from preservation for, whereas preservation can attempt only to maintain the *status quo*, conservation is more flexible in its approach and strives to improve existing conditions for the benefit of all. It seeks to reconcile divergent interests and to forge the best possible compromise. Consequently conservation finds itself at times working with, and at other times working against, the proposals of the geologically concerned industries.

The difficulties to be surmounted in arriving at a generally acceptable compromise can often be eased by a multi-purpose form of land-use. In some cases the same area of land can be used simultaneously to meet a number of needs. The mining of aggregates as practised in the United States, and currently under debate in Britain, allows normal surface activities such as farming to continue undisturbed and, through its much reduced impact on amenity as compared with open quarrying, has little adverse affect on recreational use. In other cases the uses to which an area of land can be put are consecutive rather than concurrent and the final effect can be either to restore the surface to its original use (e.g. farmland in English Midlands worked for Jurassic ironstone and later brought back to cultivation) or to put the land to a completely new use which is acceptable as a permanent feature of the landscape (e.g. farmland used for gravel winning before being developed for housing). To make the most of the opportunities inherent in such sequences, it is essential that they are planned as a whole and that the final land usage is 'determined before the first change is made; only thus can the entire sequence of operations be directed towards an agreed end. If this is not done sterilization of natural resources through premature alternative development is liable to occur, a feature now widespread in and around built-up areas. For example, lack of forward planning in south-eastern England before 1939 allowed housing estates, factories, roads, and even reservoirs to be built on land rich in surface mineral resources before these were extracted. The concept of multi-purpose land-use is not new, especially in urban areas—the headquarters of the Institute of Geological Sciences in central London stands in a worked-out nineteenth-century gravel pit. What is new, however, is the realization that the sequence of changes must be planned as an entity to ensure both the most thorough use of all resources and also the achievement of the most acceptable final state.

The search for a compromise over land-use can be brought to a successful conclusion most readily if all parties with conflicting interests in an area make contact with one another at the earliest possible stage of their decision-making; if contact is not made until one or more have crystallized their proposals into final form, room for manoeuvre is very limited and confrontation becomes almost inevitable. On the other hand, if every party is aware of the interests of the others at an early stage, or preferably before planning starts, there is scope for each to adjust its plans, as yet drawn up only in outline, to accommodate the needs of the others. Such adjustments can be made at an early stage in the development of a project, whether a conservation area or a mineral extraction site, without incurring the delays and extra expenditure that become increasingly involved as more and more effort is devoted to detailed site assessment.

2. Geology and conservation

Geological conservation is a branch of applied geology. Within conservation the science has two roles. First it must ensure that the environmental features (i.e. outcrops) which it requires for its own purposes are conserved; in this role geological facts are blended with planning considerations to achieve the desired result. In its second role geology contributes knowledge and expertise to the whole spectrum of conservation in its concern for the study and improvement of the environment. In this role geology is blended with the wide variety of other subjects, scientific and non-scientific alike, which together form the constituents of conservation.

The preceding chapters demonstrate that geological considerations influence present-day society in many ways. Society consequently requires the expertise of a practising community of earth scientists. The existence and growth of this community, and the further development of the earth sciences themselves, depend on the continued availability of a wide range of field facilities. Some are required for training the increasing numbers who study the subject, others are used to test existing hypotheses and as sources of data on which new hypotheses can be based in the prosecution of research, while still others are internationally or nationally recognized standard sections from which formations or stratigraphic stages are defined or from which type-specimens of rocks, fossils, and minerals are obtained.

Britain has played a major role in the development of the earth sciences so that many phenomena were first described from British examples and many British rock sections have come to be adopted as internationally recognized standards. The intensity of the development pressures to which the assemblage of features of high geological interest is subjected was soon recognized, and to ensure its survival Britain set up a national system of geological conservation under Government auspices on the formation of the Nature Conservancy in 1949. This lead has been followed elsewhere as the need for geological conservation has been recognized.

The second and broader role of geology in conservation stems from the inter-relations between geology and the environment. Geology investigates the factors which determine the physical environment, studies former ecosystems, and traces the development of present floras and faunas. Its involvement with time enables it to supply the historical perspective for the other natural sciences whose concern lies in the present. Geological factors thus have wide influence on the other disciplines included within conservation and have to be fully taken into account in every overall assessment of the quality of the environment.

The basic facilities geologists require in the field are relatively simple, whether they are to be used for educational, research, or reference purposes. They comprise the existence of rock outcrops, whether natural or artificial, containing sufficient interest to make their investigation worth while; convenient access to such outcrops and an assurance that each outcrop will have a reasonably permanent existence. Permanence is essential in the case of reference sections, which ideally should be preserved in perpetuity. It is also highly desirable for localities of research importance to allow future workers to use newly devised techniques and to test newly formulated hypotheses on previously studied material. It is of lesser benefit, but still desirable, for training areas.

The needs of geological conservation contrast strongly with those of the more familiar biological conservation. This contrast arises partly from differences in the intrinsic nature, distribution, and management requirements of the scientific phenomena to be conserved. In general, geological interest is resistant and can readily

313

survive threats which frequently prove harmful to biological interest such as trampling and over-intensive visiting by the general public, changes in agricultural practices, the use of toxic chemicals, and variations in the water regime. On the other hand, geological interest cannot reproduce itself and there is no possibility of recovery from damage or partial destruction. Biological interest is the more widely dispersed; there can be few areas without some potential for education and research in biology and there is a continuous spectrum from localities of low interest to areas of international importance. Many biological localities of the highest value are backed up by several alternatives which differ only slightly in quality and, in consequence, the impact of biological fieldwork can be spread over wide areas. Geological interest is restricted to a number of sharply defined areas where rocks can be seen; the bulk of the countryside has little or none. There is a general absence of duplicate localities, even if a substantial reduction in the quality of the scientific interest is accepted, and the impact of geological education and research falls on relatively few, high-grade areas. There is a concentration of geological interest in those environments most affected by man reflecting the importance of exposures which are essentially a by-product of industry. Localities of biological interest, on the other hand, are concentrated where human intervention has been least. Geological conservation is thus the more deeply involved with those industries which have provided many of its most important field facilities and on whose continued co-operation the continued existence of these facilities depends.

3. Conservation and the extractive industries

The activities of the extractive industries provide the chief meeting-ground of geology and conservation; here, as nowhere else, the consequences range from co-operation to confrontation.

Mineral extraction has been by far the largest industrial contributor to geological knowledge. Workings provide a welcome supplement to natural outcrops, especially where these are few, and also have the advantage of being fresh, extensive, and readily interpretable in three dimensions. In some, the progressive retreat of a face provides unique opportunities for the determination of rock structures and relationships. Those who work in the industry have a tradition, dating back for centuries, of drawing the attention

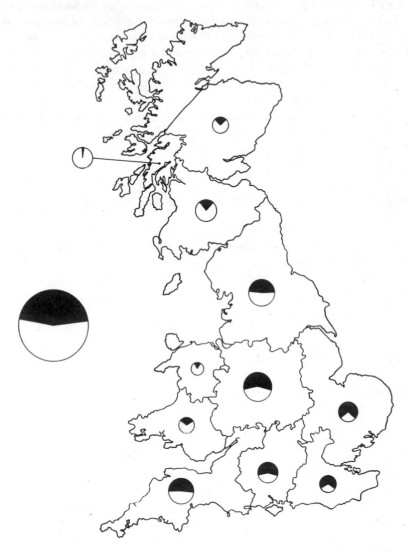

Fig. 13.1. Regional variations in the proportions of quarries among the geological sites conserved by the Nature Conservancy Council. The size of the circles is proportional to the number of sites; the size of the black segments to the number of quarries. The position for Britain as a whole is summarized on the left.

of geologists to features thought to be of scientific importance and of preserving specimens thought to be worthy of attention. National museum collections of minerals, fossils, and man-made implements owe much to co-operation by quarrymen.

Extractive sites form by far the most numerous group among localities conserved in Britain for the benefit of earth scientists, amounting to approximately 45 per cent of the 1500 total (Fig. 13.1). Their proportions differ across the country, being highest in East Anglia (78 per cent) and in south-east England (68 per cent) where natural outcrops are sparse, and lowest in north-west Scotland (3 per cent), a region characterized by an abundance of natural exposures. These estimates, when combined with the distribution of geological field education (Black 1974), suggest that quarries, active and disused, provide 25 per cent of the facilities required by the British universities for undergraduate field training in geology (Fig. 13.2). In addition they play an important role in research, though this unfortunately cannot be even approximately quantified.

There are relatively few occasions when the needs of the mineral extraction industry conflict with those of geological conservation. From time to time, however, the development of mineral workings threatens to destroy features of educational or research value, as where a quarry works rock whose interest is not evenly dispersed but occurs, for example, as highly fossiliferous pockets. There is a clear need to preserve such concentrations of interest for future study *in situ*, a need appreciated by many quarry managements who are sympathetic and willing to modify their extraction programmes to prolong the survival of the interest over short periods of time. They are, however, rarely willing or able to agree to the long-term concessions that permanent conservation would require on the grounds of the serious disruption of their programme and the considerable financial loss such concessions would involve. Similarly, material which itself is economically worthless, but has to be removed as a quarry develops, e.g. overburden or fissure fillings, can have a high scientific interest. Again, some rearrangement of working programmes may permit an extension of the time available for scientific study, but permanent conservation is once more difficult to achieve.

In general, however, such cases are rarely encountered and the conservation of the geological interest of extractive sites raises few problems while production continues; it is only when extraction ceases and the sites are put to an after-use by their owners, or come to be regarded, either officially or by the general public, as derelict land, that difficulties arise. Many after-uses are not compatible with the retention of geological interest, and disused quarries, though retaining their scientific value for many years, have little scenic appeal and are obvious targets of 'improvement' schemes which inevitably seem to require the burial of the faces in the interests of landscaping. Further, one of the conditions most frequently im-

1971-1972

TOTAL STUDENT FIELD
DAYS = 70,000

5000 student days

500 student days

Fig. 13.2. The distribution of British undergraduate field education in geology for the academic year 1971–2.

posed on operators seeking to open or expand mineral workings requires eventual reinstatement of the site. Such a condition, if strictly enforced, can only result in the total loss of any geological interest the site may have come to have during its working life. Since it is impossible to predict in the planning stage which workings will expose features of scientific value, a more flexible approach is required so that future exposures of merit can be conserved and not lost through the over-strict application of a planning condition imposed, ironically, in an effort to promote conservation.

Disused mineral workings are put to a wide variety of uses, many of them profitable to their owners. A full list of known after-uses includes tips, constructional sites of all types, water storage, restoration to agriculture or forestry, storage of bulk materials, military training, sports centres and recreation grounds, scrap yards, and a number of more esoteric uses such as animal dens in zoos. Some forms of after-use readily allow the retention of the scientific interest of the old working because they do not affect the walls but use only the floor. As long as the old faces are not landscaped and access to *bona fide* geologists is permitted, there is no objection from a conservation viewpoint for old workings being used as the sites of factories, housing schemes, or public parks. Other uses are not so favourable towards geological conservation; the use of old quarries for bulk storage or scrap yards typically involves material being piled against the old faces so that full exposure is only rarely achieved, while use for military training or for housing animals can be highly restrictive of access. After-uses such as tipping and water storage imply the loss of part or all of the scientific interest and thus are the most deleterious, though compromises can commonly be arrived at where the scientifically most important part of an old working is kept free from tip. This has worked satisfactorily in many British examples, the most noteworthy being the long-disused Barnsfield Pit at Swanscombe, which is now shared between the Swanscombe National Nature Reserve (the source of a human skull, until recently the oldest known in Europe), a large rubbish tip now soiled over, and a substantial sand storage yard.

Opposition to unsuitable after-uses of mineral workings is particularly necessary at the present time when the closure of small and medium-sized quarries is balanced, in terms of production only, by the opening of a lesser number of much larger units. Though the replacement of several small quarries by one large working is economically advantageous, the scientific interest of each new large site is less than the total interest of the scatter of smaller sites it supersedes. Consequently, if quarries now currently approaching the end of their working life are not conserved, there will be a progressive decrease in the national stock of geological assets.

Restoration of old workings in the interests of amenity also raises problems for geological conservation, though these can be overcome if the need to preserve geological interest is taken into account alongside the desire to improve the environment, especially as the desire for restoration is not motivated by financial profit. It is commonly found that it is not the old faces to which exception is taken, but the persistence of derelict and often vandalized quarry plant or the use of abandoned and uncared-for sites as illicit tips for garden refuse, household waste, and unwanted motor-cars. Old faces present a safety problem, both as a hazard to the unwary who fall from above and as a challenge to children who climb from below. Where old workings fill with water, a further hazard exists, for many such pools have one or more vertical or steeply sloping walls descending to a great depth and forming lethal traps for children. Compromises between the need to remove such hazards and to improve the environment on the one hand and the requirement for geological field facilities on the other can, in most cases, be readily reached. The flooded parts of workings are of no further geological use and can be filled with waste material to just above the water table with no loss to the scientific value. Old buildings and plant can be removed and access can be restricted to prevent illicit tipping. In a typical disused working some parts of the faces are of greater scientific interest than others, and these can be preserved while a programme of partial filling reduces the hazards presented by the other faces. Examples of such compromises are numerous and can be found on all scales. In Sunderland the very extensive Fulwell Quarry complex has been restored and partially infilled while leaving the most important scientific faces accessible; at the other end of the scale the local Naturalists Trust have cleared a very small quarry at Tasker in Shropshire of illicitly dumped material and have restricted the entrance to prevent further material accumulating.

The relationship between the mineral extractive industry and geologists is thus one of general co-operation. Conservation problems typically arise only when mineral working comes to an end. For geomorphologists, on the other hand, the relationship is not so happy and conservation problems arise at a much earlier stage, generally as soon as working starts. The operations of the extractive industry lead towards the removal of superficial landforms *in toto* and create little in return; there is no geomorphological analogue to the disused quarry faces which continue to provide geological field facilities long after working has ceased. The problem of landform conservation is most acute in two fields: the destruction of entire landforms of fluvio-glacial origin in the supply of sand and gravel and, in greater complexity, in the working of limestone.

Many limestone surfaces have been weathered into pavements which are of great interest to the geomorphologist and constitute a habitat of much value to biologists. Unfortunately such surfaces are exploited for ornamental stone, largely for garden rockeries, and since only their uppermost layer, most sculpted by the weather, is saleable, the production of relatively small quantities of stone entails the destruction of the pavement and its scientific interest over wide areas. This form of mineral production, completely negligible in tonnage produced, is one of the most destructive forms of the mineral extraction industry in terms of conservation. Deep quarrying in limestone also raises conservation problems through the damage and destruction done to cave systems. These are of very considerable scientific interest: they are the only place where habitats unaffected by man may still be found in long-settled countries; they contain a unique record of the development of present faunas; their study provides irreplaceable evidence of the development of drainage systems; and they are of practical use for public water supply, recreation, and the interest they generate in the lay public. The problem of ensuring the conservation of the most important cave systems, while not interfering unduly with limestone production, is currently acute in many countries. No piecemeal solution to this problem has any prospect of success and the clash between mineral extraction and conservation interests can be resolved only by the evolution of national mineral-working policies. Such an approach, which could profitably be adopted also in the extraction of sand and gravel, could identify and conserve the scientifically most important areas, while encouraging the exploitation of resources elsewhere in accordance with prediction of national needs.

To biologists the operations of the mineral extraction industry afford both advantages and disadvantages which vary both in character and severity during the exploration and exploitation of a mineral deposit.

Exploration, which in its initial stages characteristically consists of geological mapping and sampling with related geochemical and geophysical surveys, requires only occasional visits by individuals or small parties and therefore has little, if any, influence on the biological interest of an area. Even where drilling takes place, the impact can be insignificant, though the erection, use, and subsequent removal of drilling equipment, the passage of supporting vehicles, and the construction of temporary buildings and other small-scale works can have a harmful effect on areas of localized or delicate interest. The more intensively an area is explored in this way, the more serious the damage likely to result; where boreholes are sunk in a closely spaced pattern they can have a serious adverse effect even on quite robust biological phenomena.

Mineral extraction is much more severe: it involves the destruction of the existing environment on a scale determined by the size of the deposit being worked and by the extraction methods employed. Open pits and quarries are the most destructive method, for within their final limits nothing of the original interest will survive (Zuckerman 1972). The impact of extraction, moreover, is not confined to the extractive site itself but spreads beyond it over considerable tracts of adjoining land. Spoil heaps bury the original land surface. Water pollution by heavy metals leached from newly exposed country-rock, rock waste, or tailings spreads along surface or subterranean drainage systems, often for many miles, and can have effects which outlast the extraction activity by many years. Pollution from dust has an adverse effect and may alter entire ecosystems when the dust is biologically active. Noise and ground vibrations, either from blasting or machinery, disturb wildlife.

Of the various factors which control the amount of damage done to biological interest, the size of the extractive site is by far the most important. The greater the yield and value of mineral extracted per unit surface area, the less the resulting destruction of scientific interest. There is a strong contrast between the conservation attitude to workings which yield only minimal tonnages but destroy scientific interest over a widespread area as compared with well-designed excavations with high faces which yield large tonnages of nationally important minerals per hectare of surface consumed.

The benefits of mineral extraction to the biological interest of the environment accrue almost exclusively when workings become disused. They stem mainly from the creation of new types of habitat in areas where they were formerly absent or rare. One commonly encountered example is the creation of open water and marshes by the flooding of surface excavations or subsidence above old mine workings; another is the production of artificial cliffs and screes through quarrying and the accumulation of mine waste. Additionally, abandoned workings provide derelict ground which can be as scientifically rewarding to the biologist as to the geologist, especially as abandoned workings gain an additional interest from the artificial habitats provided by the old buildings and other works in and around extractive sites. The ecological effects of mineral exploitation in the U.K. and their significance to nature conservation have been summarized by Ratcliffe (1974), who states: 'It is true to say that while mineral exploitation has caused some damage to nature conservation interest in Britain, this has been mostly on a local and minor scale, and on the whole the gains have outweighed the losses, especially in the creation of interesting new habitats. By comparison, the impact of agriculture on wildlife has been infinitely more destructive.'

The after-use of old mineral workings can, as in the case of workings of geological significance, have either a beneficial or an adverse effect on the scientific interest they have acquired. Some surface workings are put to after-uses which preserve the new type of habitat extraction has created; the biological interest can then be perpetuated. Old gravel workings landscaped into permanent lakes retain the new habitat and its value to conservation can be increased if a portion is reserved as a wildfowl sanctuary or as an educational area. In other cases, the after-use of an extractive site destroys the new habitats and their scientific interest by leading to a drastic change in conditions. For example, the filling of old workings by tipped material can only destroy their new habitats so that the eventual restoration of the tip surface for productive use, such as farming, is achieved only at the cost of scientific loss which in some cases may exceed the advantages gained in the restoration. Old workings used for the disposal of toxic wastes present a severe problem: the tipped material can pollute wide areas should the tip accidentally leak. The dumping of toxic waste in areas of slow-moving or complicated drainage pattern is especially hazardous as the pollutants can linger and spread more widely.

The diversity of roles played by a single major mineral extractor is well exemplified by the London Brick Company, the chief producer of bricks in Britain. At present the Company has 27 pits, of which 21 are active, with a total worked area now approaching 2000 hectares and increasing at the rate of some 40 hectares each year. The pits work the Oxford Clay along a considerable length of its outcrop in the south and east Midlands of England and provide the only significant exposures of this formation in their area. Individual pits cover areas as large as 400 hectares and have faces in excess of 50 metres in height.

The Company's properties are visited by over 300 geologists each year. For those wishing to do research for a higher degree, special arrangements for access are made; for the more casual visitors, parties are organized in conjunction with the Geologists' Association to minimize disturbance to the Company's normal activities. Individual visitors are not catered for but are put in touch with the Geologists' Association to await the next visiting party. In the past the Company has supplied fossils for educational use and, more recently, has provided school examination boards with matched sets of closely similar specimens.

In the past ten years the skeletons of five large marine dinosaurs (Fig. 13.3) have been extracted from the pits. Four of these specimens have been given to the British Museum (Natural History); they include a specimen of *Pleurodon ferox* which ranks as one of the largest ever discovered. The extraction of two of these specimens has been featured on television and involved the rearrangement of the working programme of the pit.

The pits are put to a wide variety of after-uses including amenity, fishing, angling, water-skiing, and waste disposal. One pit has found an unusual use in that it has been used for an inland diving training school, many of whose graduates are now working on North Sea oil projects. The Company has always been conscious of the effects on the environment resulting from the extraction of clay for bricks, and for many years the Company's Estates Department has been reclaiming and improving the land from which their raw material has been taken.

Fig. 13.3. The skeleton of a plesiosaur belonging to the genus *Cryptocleidus* discovered during extraction of Oxford Clay near Peterborough. (Photograph: London Brick Company.)

For some time now certain local authorities have had the facility of using the Company's clay pits as outlets for their refuse. In 1970 eight authorities tipped a total of over 60 000 tons of refuse into London Brick's pits in the Calvert, Bletchley, Bedford, and Peterborough areas. Arrangements such as these offer outlets to authorities' collection vehicles enabling them to tip directly into the Company's pits, which are operated by the local council using controlled tipping principles.

Since 1963, a project for filling pits in the Peterborough area has been in operation in conjunction with the Central Electricity Generating Board and local planning authorities. The material used is fly ash, the dry waste residue from pulverized coal burned in electricity generating stations on the River Trent in Nottinghamshire. It is received in liner trains, similar to cement wagons, at the rate of 4000 to 6000 tons per day on a five-days-a-week basis. At the terminal the fly ash is discharged using compressed air, mixed with water, and pumped through pipelines into the pits, where it settles. The water is then decanted and recirculated for further use. After a pit has been refilled the surface is covered with topsoil and is returned to agricultural or other uses. The project has been established by the CEGB and in the next 20 or 30 years it is anticipated that over 1200 ha will be treated in this way (Fig. 13.4).

In 1970 the London Brick Company commissioned a market survey and feasibility study on refuse disposal. The study showed that about one-third of the country's population is within 80 km of London Brick's pits, which are located along a line drawn roughly between Oxford and the Wash. Such an area, which includes the north London conurbation, produces 7 million tons of refuse each year. Within an 80 km radius it is considered economic to transport refuse to the pits by road. Longer distances are served more economically by rail.

A subsidiary company, London Brick Land Development Ltd., offers local authorities a package deal to solve the most difficult problems of refuse disposal which, when operated within the area specified above, is cheaper than incineration. The essence of the scheme is that London Brick Land Development erects and operates transfer stations on urban sites provided by local authorities. Refuse will be received at these specially designed stations from local collection vehicles and mechanically compacted and pushed into 2·5 m × 2·5 m × 6 m totally enclosed steel containers by hydraulic rams. In this way low-density refuse can be compressed into economic payloads. When a container is fully charged it is loaded on to a bulk haulage vehicle for transport to the pit terminal. Similar systems are currently operating in America, Holland, and Japan. On arrival at the pit terminal the container is transferred to a specially designed slave vehicle which discharges the refuse at the bottom of the pit in accordance with scientifically controlled tipping procedures. The refuse is covered each day with a layer of earth or other suitable material to eliminate the problems of smell and infestation.

Many other proposals are under consideration. The use of pits for water storage, for golf courses, and for industrial development is being reviewed and proposals have been put forward for the use of a restored pit as a leisure park in association with the Duke of Bedford and Chipperfields. Restoration schemes have now been decided for more than half of the total area of the Company's pits and future restoration schemes affecting 800 ha of land still to be dug for clay have been decided upon.

The London Brick Company thus has a wide spectrum of activities beyond the manufacture of bricks; some are of great importance to geological conservation, while others aim at improving the environment as a whole.

4. Conservation and the construction industries

The recognition of a canal engineer, William Smith, as 'the Father of Geology' acknowledges the debt the development of geology as a science owes to the construction industries. Since Smith's day geologists have continued to draw much information, not obtainable elsewhere, from construction sites.

Excavations in connection with the transport industries have attracted protracted attention as cuttings for canals were succeeded, first by those for railways, and later, by those of the motorway programme. Cuttings are of particular value to the geologist in that they supply natural traverses with continuity of exposure, up to several miles in length, through strata which otherwise are only discontinuously exposed.

Tunnels are similar in continuity of exposure to cuttings but are more difficult of access and less convenient for study. Once again the purposes of construction have changed—from canals and railways to roads and aqueducts—each system providing its own opportunities for scientific work. Although most exposures in tunnels are open only temporarily, they have the great advantage of being divorced from the land surface and thus demonstrating the vertical dimension of rock structure and disposition to a unique extent. Further, as the services of geologists have long been employed in predicting the material through which tunnellers must cut, tunnels have acted as a stimulus for the development of geological techniques. The construction of a major tunnel in an area of complicated geology constitutes a severe proving ground for hypotheses framed in the course of academic research and tests the reputation of the consulting geologists involved.

Excavations for foundations have the same advantages for geology as mineral workings. They can be regarded as a peculiarly

Fig. 13.4. A sequence of photographs taken from the same viewpoint in 1963, 1969, and 1973 showing the progressive reclamation of a pit in the Oxford Clay near Peterborough by infilling by fly ash. (Photographs: London Brick Company.)

temporary form of the latter, whose after-use is definitely fixed in advance so that no compromise with conservation interests can be considered. They are geologically significant through providing exposures in areas which typically are devoid or poor in natural outcrops; since urban areas tend to be situated on flat, outcrop-free ground, the foundation excavations are often the chief contributor to our knowledge of urban geology.

A further branch of the construction industry, the laying of pipelines for oil, water, and gas, presents opportunities to geologists which are unusually ephemeral. The trenches in which pipelines are laid are rarely open for more than very short times and, even if the route of the line is known, there is the chance that the excavation will not reach bedrock where desired. Some use, however, is made of pipe trenches for scientific purposes, despite the difficulties involved.

For geological conservation, the construction industry comes second only to the mineral extraction industry as a supplier of information. The exposures it provides present conservation problems which are less varied but more intractable than those associated with mineral working: most of the exposures are very temporary and, no matter how important they may be scientifically, their long-term conservation is impracticable. It is accordingly necessary to arrange the most thorough possible recording and sampling of the exposures as they become available. Unfortunately the organization of a national network of recorders for all temporary excavations has never been achieved.

One advantage construction sites of all types possess is their ability to supply specimens in exceptionally large quantity and of exceptionally good quality through being unweathered. During the construction of the M 4 on the eastern bank of the Severn, for instance, the cuttings on the approach to Aust Bridge exposed the important Rhaetic Bone Bed over a large area. By arrangement with the contractors, facilities were granted whereby several tons of this material, normally available only in small quantities, were collected and stored for future study.

Where a deposit of unusual scientific interest extends, but does not crop out, over a moderately wide area the activities of the construction industry provide the only possibilities for its study. If development is sufficiently intense, a sequence of temporary outcrops can be produced so that there is effectively continual but interrupted exposure. Through arrangements made with local planning authorities, notice can be given of all excavations of the requisite depth within the outcrop so that at most times there will be a section available for study. Arrangements of this sort have been made at Crayford in Kent, where the well-known Brickearth has no permanent outcrop but is exposed repeatedly by temporary excavations.

Cuttings can provide permanent exposures of the harder rocks they traverse. Unfortunately many engineers regard rock exposures in cutting walls as undesirable and in the interests of safety or amenity batter them back, soil them over and seed them with varying degrees of success. It is questionable whether an artificially bevelled slope of this sort constitutes an improvement in amenity; what is certain is that it constitutes an unnecessary loss of a geological asset. Fortunately the fashion for battering cutting walls appears to be on the wane, a development to be encouraged in the interests of conservation.

With all exposures created by the construction industry, access can present problems. Visits by geologists require forward planning if they are not to disrupt work schedules, and it is important that engineers and contractors accept the need for, and encourage, such visits, for the information collected may well prove of great use when construction moves to a near-by site. Access to railways and motorways is also important. Since pedestrians are discouraged by law from motorways in Britain, arrangements whereby *bona fide* research geologists can visit motorway exposures for scientific purposes have had to be evolved.

For biological conservation, the construction industry presents problems very similar to those posed by mineral extraction: in both cases the land immediately affected loses all its scientific interest. On the other hand, the activities of the construction industry, especially where transport routes are concerned, bring some compensating advantages. For instance, the verges and central reservations of motorways are not affected by agricultural practices and are not liable to further development and come to form a refuge for the natural flora and fauna of the surrounding district. The construction of canals and reservoirs has provided wet-land habitats where formerly there were none; and all forms of disused transport routes rapidly become colonized by wildlife and acquire a great interest to naturalists.

Geology in conservation

In general, the effects of the construction industry are much more localized than those of the mineral extraction industry, especially as they do not involve pollution. The construction of reservoirs stands apart, however, for, although the construction of the dam itself makes relatively little impact, the flooding of the valley upstream affects biological interest over a wide area through submergence and alterations in the water regime above top water level. As is typical with the impact of the construction industry on scientific interest, little can be done to mitigate these effects other then a rescue investigation, perhaps financed by the future reservoir owners, before destruction occurs.

5. Conservation in Britain

In Britain the Nature Conservancy Council is the official body established by Act of Parliament in 1973 to be responsible for the conservation of flora, fauna, and geological and physiographical features. It is financed by the Department of the Environment but is free to express independent views. It establishes, maintains, and manages National Nature Reserves and has the duty of advising the Government on conservation policies and on how other policies may affect nature conservation. It is a source of advice and knowledge for all whose activities affect the natural environment and it commissions, supports, and undertakes relevant research. The Geology and Physiography Section discharges the responsibilities of the Nature Conservancy Council in the earth sciences and has the following functions:

(1) Conservation of those British localities of geological and physiographic interest which are essential to the continued study of these sciences.
(2) Provision of advice, largely based on a combination of conservation and of geological and physiographic factors, both within and outside the Nature Conservancy Council.
(3) Liaison with all persons with an interest in the geology and physiography of Britain.

The first two functions follow from the 1973 Act, and the third is a prerequisite in fulfilling the others.

The Council has declared 151 National Nature Reserves which cover nearly 120 000 ha. Some are owned or leased; others are established under a nature reserve agreement by which the owner or occupier, or both, retain their rights but agree to the management of the area in a way which fully takes into account the interests of nature conservation. Owing to the durable nature of the majority of geological sections and landforms, little need has been found to give them Reserve status. Only seven areas, all exceptional, have so far been declared. Some thirty other Reserves, declared for biological reasons, are of high interest to earth scientists. Of the seven, Swanscombe is part of a disused gravel pit, which also houses a council refuse tip and a sand stockpile, in an urban setting in north Kent. It has been the scene of very extensive geological research and was originally declared as the locality where the oldest known human skull from Europe was discovered in association with numerous animal remains and flint implements. Since it became a Reserve, further finds have been made by the Royal Anthropological Institute in strata below the skull-bearing horizon where, once again, animal and human relics have been found in association. The declaration of the area has not only prevented an unsuitable after-use, but has given security of tenure to the investigators, who have been enabled to carry out a long-term research programme. In consequence of the security, results of importance have been achieved and the scientific value of the site further enhanced.

Wren's Nest is the geological Reserve most used for education. A nature trail and associated facilities, including a reserve handbook and, through co-operation with the local authority, a museum display, have been provided. The trail attracts several thousand visitors at all academic levels each year, and through its design they are led away from the most important research localities where serious damage could occur. The reserve occupies an area, once derelict land, from which limestone was extracted for at least 500 years.

The classic locality of Knockan Cliff, within Inverpolly National Nature Reserve, was crucial in the controversy between Murchison, Nicol, Geikie, and Lapworth a century ago, a controversy which terminated with the recognition for the first time of a major tectonic thrust. A nature trail has been provided for the benefit of visitors and the area is much used for educational purposes (Fig. 13.5). Glen Roy Reserve is of international interest because of the well-developed beaches cut by ice dammed lakes during the recent deglaciation of the area. These 'Parallel Roads' have been in-

Fig. 13.5. A 'geological wall' erected on the Inverpolly National Nature Reserve at the start of the Knockan geological nature trail. The specimens reproduce the stratigraphy of the Reserve in miniature and help visitors with little or no geological knowledge to identify rocks on the trail.

vestigated since the eighteenth century but much research potential still remains. The Ogof Ffynnon Ddu Reserve covers part of the largest and most scientifically important British cave system. Axmouth–Lyme Regis contains the largest landslip in England. Fyfield Down has been declared to protect one of the best areas of sarsen stones still remaining. All these Reserves have considerable potential for research, and the last-mentioned is being developed additionally for educational use.

In addition to its National Nature Reserves, the Nature Conservancy Council have notified some four thousand Sites of Special Scientific Interest to local planning authorities as well as to those who own and manage them. The Council has no rights over these areas, but planning authorities must consult it before granting permission for development. Approximately 1500 sites have been notified for geological or physiographic reasons and these show great variety in character and size. Geological sites include small natural outcrops, clay and gravel pits, mines and quarries, both active and disused, stream sections, road, rail, and canal cuttings, sea cliffs and foreshores, and large tracts of rocky upland. Examples of physiographic sites are erratic boulders, springs, Chalk coombes and their deposits, raised beaches, overflow channels, eskers and kames, periglacially patterned ground, sand-dunes, spits and tombolos, caves and landslips.

The conservation of Sites of Special Scientific Interest depends on the influence the Conservancy can exert in an advisory capacity when consulted over proposed developments such as building, mineral extraction, road construction, and tipping. When such proposals affecting an SSSI are received, their implications are carefully considered. Some proposals would increase the scientific interest and receive support; others would have little significant effect and no comment is made; those which would prove adverse to the scientific interest are the subject of further action. Frequently some compromise, whereby the scientific interest is left intact through a modification of the proposals, can be reached through discussions involving the intending developers, the site owner, the local planning authority, the scientists who use the site, and the N.C.C. The Council, through the Geology and Physiography Section, draws on its nationwide experience of analogous compromises and calls in its liaison contacts with relevant expertise. A satisfactory compromise cannot be reached in a small number of cases and the Conservancy then advises the local planning authority to refuse permission.

A refusal is often followed by a planning appeal to central government by the developer and leads to a public inquiry before a Department of the Environment inspector. At such an inquiry the Section prepares and presents the scientific case, drawing witnesses from its own members and the ranks of its liaison contacts. Support is also obtained from the scientific community, both in Britain and overseas. A significant proportion of inquiries, which are expensive to all parties in time and effort, arise solely from a lack of early consultation between industrial and conservation interests. It would be to the advantage of both parties to avoid such confrontations; perhaps an increasing familiarity with conservation among industrial geologists will help towards achieving this end.

References

BLACK, G. P. (1974). The role of extractive sites in geological education and research. *Proc. R. Soc.* A **339**, 389–94.

Geology in conservation

RATCLIFFE, D. A. (1974). Ecological effects of mineral exploitation in the United Kingdom and their significance to nature conservation. *Proc. R. Soc.* A **339**, 355–72.

SMITH, P. J. (1975). *The politics of physical resources.* Penguin Books, London.

ZUCKERMAN, S. (1972). *Report of the Commission on Mining and the Environment.* Commission on Mining and the Environment, London.

Further reading

General

In offering a list of further reading it is necessary to be conscious of the needs of the individual reader and the range of topics covered in this book. The various authors have provided text references to their own contributions but it is inevitable that these cannot be comprehensive; they were, however, invited to propose items for supplementary reading. There is little doubt that some of the most readable, attractively-presented and up-to-date reading is provided by 'the publications of the Open University, including the following:

The Open University. (1974). *The Earth's physical resources. Science: A Second Level Course.*

      1 *Resources and systems.*
      2 *Energy resources.*
      3 *Mineral deposits.*
      4 *Constructional and other bulk materials.*
      5 *Water resources.*
      6 *Implications: Limits to growth.*

The Open University. (1975). *Environmental control and public health. Technology: A Second Level and Post-experience Course.*

      3 *Water: Origins and demand.*
      4 *Conservation and abstraction.*
      9 *Municipal refuse disposal.*
     10 *Toxic wastes.*

The Open University. (1976). *Earth science topics and methods. Science: A Third Level Course.*

      Urban geology case study.
      Porphyry copper case study.
      Sedimentary basin case study.
      Techniques handbook.

Adding a single book to this list, one would suggest Skinner, B. J. (1969). *Earth resources.* Prentice-Hall, New York.

Specific topics

Oil and natural gas

Hobson, G. D., and Tiratsoo, E. N. (1975). *Introduction to petroleum geology.* Scientific Press, London.

Further reading

Landes, K. K. (1959). *Petroleum geology* (2nd edtn). John Wiley, New York.

Levorsen, A. I. (1967). *Geology of petroleum* (2nd edtn). Freeman, London.

Royal Society, (1974). Energy in the 1980s. *Phil. Trans. R. Soc.*, A**276**, 405–615.

Tiratsoo, E. N. (1972). *Natural gas* (2nd edtn). Scientific Press, London.

Coal

Raistrick, A., and Marshall, C. E. (1939). *The nature and origin of coal seams*. English Universities Press, London.

Stutzer, O. (1940). *Geology of coal*. University of Chicago Press.

Williamson, I. A. (1967). *Coal mining geology*. Clarendon Press, Oxford.

Nuclear fuels

Nininger, R. D. (1954). *Minerals for atomic energy*. Van Nostrand, London.

Mining and ore geology

Bateman, A. M. (1950). *Economic Mineral Deposits*. John Wiley, New York.

Flawn, P. T. (1966). *Mineral resources*. Rand McNally, London.

Jones, M. J. (ed.), (1969). Mining and petroleum geology. *Proc. 9th Commonw. Min. Metall. Congr.* **2**.

Kreiter, V. M. (1968). *Geological prospecting and exploration*. Mir Publishers.

Lawrence, L. J. (ed.), (1965). Exploration and mining geology. *Proc. 8th Commonw. Min. Metall. Congr.* **2**.

McKinstry, H. E. (1948). *Mining geology*. Prentice-Hall, New York.

Park, C. F., and MacDiarmid, R. A. (1964). *Ore deposits*. Freeman, London.

Stanton, R. L. (1972). *Ore petrology*. McGraw-Hill, London.

Warren, K. (1973). *Mineral resources*. Penguin Books, London.

Industrial minerals

Bates, R. L. (1969). *Geology of industrial minerals*. Dover, New York.

Bouchert, H., and Muir, R. O. (1964). *Salt deposits*. Van Nostrand, London.

Gillson, J. L. *et al.* (ed.), (1960). *Industrial minerals and rocks* (2nd edtn). Amer. Inst. Min. Met. Pet. Engrs.

Johnstone, S. J., and Johnstone, M. G. (1961). *Minerals for the chemical and allied industries*. Chapman and Hall, London.

Lamey, C. A. (1966). *Metallic and industrial mineral deposits*. McGraw-Hill, New York.

Construction materials

Beaver, S. H. (1968). *The geology of sand and gravel*. Sand & Gravel Association.

Further reading

Allen Howe, J. (1910). *The geology of building stones.* Edward Arnold, London.
Knight, B. H. and Knight, R. G. (1958). *Builders materials.* Edward Arnold, London.

Water

Davis, S. N., and DeWeist, R. J. M. (1966). *Hydrogeology.* John Wiley, London.
Rodda, J. C., Downing, R. A., and Law, F. M. (1976). *Systematic hydrology.* Newnes-Butterworths, London.
Smith, K. (1972). *Water in Britain.* Macmillan, London.
Todd, D. K. (1959). *Ground water hydrology.* John Wiley, London.
Walton, W. C. (1970). *Groundwater resource evaluation.* McGraw-Hill, New York.
DeWeist, R. J. M. (1967). *Geohydrology.* John Wiley, London.

Construction

Flawn, P. T. (1970). *Environmental geology.* Harper & Row, London.
Krynine, D. P., and Judd, W. R. (1957). *Principles of engineering geology and geotechnics.* McGraw-Hill, New York.
Leggett, R. F. (1962). *Geology and engineering* (2nd edtn). McGraw-Hill, New York.
—— (1973). *Cities and geology.* McGraw-Hill, New York.
Paige, S. (ed.), (1950). Application of geology in engineering practice. Berkey Volume. *Mem. geol. Soc. Am.*
Tank, R. W. (1973). *Focus on environmental geology.* Oxford University Press, New York.

Geological hazards

Bolt, B. A., Horn, W. L., Macdonald, G. A., and Scott, R. F. (1975). *Geological hazards.* Springer-Verlag, Berlin.
Scheidegger, A. E. (1975). *Physical aspects of natural catastrophes.* Elsevier, Amsterdam.

Conservation, planning, and environmental matters

Blunden, J. (1975). *The mineral resources of Britain: A study in exploitation and planning.* Hutchinson, London.
Haywood, S. M. (1974). *Quarries and the landscape.* British Quarrying and Slag Federation.
Jones, M. J. (ed.), (1975). *Minerals and the environment.* Institution of Mining and Metallurgy.
Royal Society, (1974). Discussion on the exploitation of British mineral resources (other than coal and hydrocarbons) in relation to countryside conservation. *Proc. R. Soc.* A **339**, 271–416.
Warren, P. T. (1970). *Geological aspects of planning and development in Northern England.* Yorkshire Geological Society.
Zuckerman, S. (1972). *Report of the Commission on Mining and the Environment.* London.

Further reading

Central to the thoughts of many may be two questions ... how will it all influence me? and what really is the world resource position? In conclusion, therefore, here are two references which will not necessarily answer these questions to the full but may stimulate thinking.

Smith, P. J. (1975). *The politics of physical resources*. Penguin Books, London.

Murdoch, W. W. (1971). *Environment: Resources, Politics and Society*. Sinauer, New York. Particularly Chapter 4: Preston Cloud, Mineral Resources in Fact and Fancy; Chapter 5: M. King Hubbert, Energy Resources; Chapter 7: T. E. A. van Hylckama, Water Resources.

Index

Aberfan, 302
Afghanistan, 237
aggregates, 94, 166, 175–80
    deleterious, 170–71
    shrinking, 174
air photographs, 107
airborne geophysics, 113–15, 126, 129
alite, 198
alkalis, 217
    in cement, 171
Alston, 141, 151, 154, 158
alumina ratio, 203, 205–7, 217
Alyeska Pipeline, 28
anthracite, 66
apatite, 212
aquifer, 104, 225, 226, 230, 232, 235, 239–42,
    248, 251, 255, 290
    artesian, 226, 227
    confined, 225, 226, 228, 241, 242
    unconfined, 225, 226, 242
aqueduct, 272
area selection, ore deposits, 123–35
arid zones, 80, 129, 131, 250
armouring stone, 183
aromatics, 37
Arsenal Well, Denver, 250, 299
artesian flow, 40, 46, 258
artificial recharge, 243, 254
asbestos, 78, 93, 104, 108
Asmari Limestone, 35
Aust Bridge, 328
Australia, 129, 131, 212, 256
Axmouth–Lyme Regis landslips, 332

Barnsfield Pit, Swanscombe, 318
barytes, 141, 144, 158
bauxite, 180, 212–13
Bahamas, 212
belite, 198
Bermuda, 212
bicarbonate, 246–7
bill of quantities, 271

biological aspects of conservation, 314, 320–22,
    329
borehole cores, 134–5, 214, 232, 235, 267–8,
    278
bornite, 99
Bougainville, 81
Brazil, 282
breakwater, 183–4
brick clays, 94, 189–91
brickearth, 329
brine, 5, 35, 40, 162, 248, 301
Britain, metallogeny of, 137–65
British Gas Corporation, 31
British National Oil Corporation, 53
British Petroleum Company, 48, 51
brucite, 215
building stones, 191–4
bush fires, 304

calcium orthosilicate, 198
Caledonian, 140, 141, 146, 154, 158
California, 244, 251
Canada, 113, 123, 125–7, 129, 131, 292
canal, 327, 329
Canary Islands, 237
carbonates, 34, 41, 246
carbonatite, 212
Carboniferous, 144–5, 207, 210, 220–1
casing, borehole, 237, 239
cassiterite, 159
caves, 320
Central Electricity Generating Board, 324
cement, 196–223
    –aggregate reaction, 171–2
    blending raw materials, 196, 200–3, 205, 206,
        218–21
    clinker, 198–200, 205, 207, 215, 217
    kiln, 200–1
Chad, 256
chalk, 228, 241, 251–2
chalcedony, 171
chalcocite, 92, 99

Index

chalcopyrite, 85, 92, 99, 157, 158
chert, 170, 218
Chile, 82
china clay waste, 187–8
chromite, 140
chrysotile, 102
civil engineering, 83
claims, 283–4
clays, 167, 170, 172, 179, 207, 218
cobalt, 88, 90, 92, 98
coal, 2–3, 9, 38, 54–77, 170, 172, 174
  cannel, 57
  in China, 10–11
  reserve evaluation, 19–20, 61–3, 76
  seams, 56–8
  UK production, 56
  waste, 187, 302
coalfields
  Durham, 74
  E. Midlands, 58
  Lancashire, 57
  Mid-Lothian, 54
  Oxfordshire, 63
  Pembrokeshire, 76
  Selby, 11, 63
  Somerset, 54
  South Wales, 66, 75, 302
concrete expansion, 217
cone of depression, 239–41
connate water, 225
conservation, 310–38
construction industry, 3, 327–30
  stone, 181
contract documents, 260, 280
copper, 84, 88, 98, 102, 107, 141, 144
coral limestone, 212
Cornwall and Devon, 144, 148, 150, 151
Cretaceous, 207, 220, 256
Cromagnon Man, 2
cross-course, 155
crude oil, 10, 38, 39
crustal pressure, 40–1, 295–8
cuttings, 327

dams, 275–7, 330
dam failures, 285–6
  investigations, 262
dams and reservoirs
  Balderhead, 286, 299, 301
  Farahnaz Pahlavi, 275–7
  Glen Canyon, 172
  Koyna, 297
  Llyn Brianne, 183
  Malpasset, 285
  Mangla, 263–4
  Monteynard, 297
  Mornos, 272–3
  Nurik, 183
  Nyumba ya Mungu, 182
  Roseires, 281, 283–4

  Scammonden, 183
  Thames barrage, 303
  Vaiont, 280, 306
  Warragamba, 285–6
Darcey's Law, 299, 234, 235, 251
Deccan Plateau, 297
demand for minerals, 81
Denver, Colorado, 155, 250, 301–2
Department of the Environment, 330, 333
derelict land, 66
derrick stone, 183
dewatering, 242, 303
diamonds, 79
diatomite, 174
dinosaurs, 323
disused workings, 318
dolomotization, 35
Dortmund, 254
downhole logging, 108
dredged aggregates, 195
drill-core, 88, 109
drill-holes, 90–1, 100–1
drilling, 46, 48, 63–4, 69, 96, 106, 132–4,
  213–14, 265–9, 321
dry process for cement, 201, 206, 210
Dunbar, 203, 222

earthquakes, 5, 232, 289, 291, 297
education, 33, 107–8, 110
effective stress, 298
employment, 261
engineering geology, 94, 98
environment, 26–30, 81
ERTS, 8, 26–7
Eocene, 288
evaporation, 233–4
excavation, 208, 210, 269–70
exploration, 7–14, 56, 61, 63–4, 84–92, 98,
  111–36, 321
extractive industries, 314–17

faults and faulting, 44, 59, 75, 96, 103, 154, 289,
  296, 300
ferrite phase, 198
field work, 316–17
fill, 94, 106
Finlayson, A. M., 138–48
fireclay, 58, 74
floods, 293–4
flotation, 92
fluid injection, 298
fluids, mineralizing, 155
fluid pressures, 155–6
fluorite, 158, 217
folds and folding, 88, 96, 97, 104, 141, 149, 289
formation waters, 161
fossils, 313, 316
foundations, 273–5, 303
fractures, 35, 125, 149, 150–6, 160

Fullwell Quarry, Sunderland, 319
furnace bottom ash, 185

gas, 18, 60, 277, 296
Gault Clay, 208
Geneva Convention of the Sea Bed, 25, 51
Germany, 254
geochemistry, 95, 99, 100, 108, 120–3, 132
geologist, role of, 13–14, 47–9, 60–1, 65, 80,
    214–15, 271, 280–1, 313
Geologists' Association, 323
geomorphology, 320
geophysics, 6–9, 47, 70, 106–8, 113–20, 235,
    264–5, 321
  down-hole logging, 64, 70, 235
geostatistics, 104, 108, 220
geotechnical mapping, 263
geothermal energy, 51, 12–13, 162, 226
  gradient, 38
Ghyben–Herzberg relationship, 243
glass sand, 173
  waste, 186
glacial erratics, 74
Glen Roy, 331
gold, 78, 141
gossan, 85
government surveys, 84, 124
grade, 85, 89–90, 102
granite, 140, 141, 144, 158, 160, 162, 187, 194
gravel, 311
Great Artesian Basin, 256
Greece, 273
Great Ouse, 252
ground geophysics, 116–19, 130
ground pressures, 58
groundwater, 1, 3, 78, 95, 170, 224–57, 268,
    273, 277, 279, 281, 302–4
  age, 256–7
  chemistry, 245–50
  development, 250–7
  flow, 230
  levels, 230
  minerals, 78
  storage, 232
gypsum, 186, 193, 196, 198, 212, 215

Hawaii, 237
hazards, 272, 279, 281, 285–6, 287–307, 319
heat flow, 13, 160
hematite, 145
Hercynian, 140, 141, 144–5
hot waters, 160
hydraulic gradient, 229, 243
hydrocarbons, 5, 35 6, 225
hydro-electricity, 11, 29, 95, 303
hydrological cycle, 225
hydrothermal mineralization, 141, 144–5

ignimbrite, 305
illite, 171, 191, 206

Indus Valley, 255
industrial minerals, 3
Inglewood anticline, 299
injection wells, 301
Institute of Geological Sciences, 54, 67, 311
investigation, 67–8, 255, 261–71
Iran, 80, 237, 275
Ireland, mining, 79
iron, 102–3, 212–13
ironstone, 145, 311
isotopic dating, 146–9
Israel, 255–6
Ivigtut, 147

JOIDES, 9
jointing, 59–60
Jurassic, 43–4, 256, 311
  coal, 54
juvenile water, 225

kaolinite, 187, 191
Kielder Tunnel Scheme, 277–80
Kilembe, 83–107
kiln, 201, 203, 218
Knockan Cliff, 331

landfill, 250
landforms, 250
landslides, 94, 99, 105, 276, 281–2, 291, 294,
    302–3, 306, 332
laterite, 79
Launay, Louis de, 138
laurvikite, 194
laws, 24–6
leachate, 248, 250
lead, lead-zinc, 141, 144–5, 158, 185–6
Liège Metro, 284
lime saturation factor, 203, 205–7, 217, 219
limestone, 196–7, 210, 320
linnaeite, 92, 99
liquid inclusions, 156
London Basin, 241, 252
London Brick Company, 323, 325, 377
Lower Palaeozoic, 146, 154

magnesia, 215
magnetic investigation, 264
Malaysia, 212
Man and environment, 4, 6, 310
mapping, 89–90, 91, 93, 95, 125, 263–4
marcasite, 172
Mediterranean, 256
Mesozoic, 44–5, 145–6
metallogenetic (mineral) provinces), 125, 138–
    145
metamorphic rocks, 104, 125
metasediment, 88, 96
meteoric waters, 225, 248
methane, 37, 38, 60, 290
Mexico, 212

Index

Middle East, 237
migration of oil and gas, 44
mines
  Avoca, 140
  Bancroft, 104
  Boulby, 96
  Butte, Montana, 93
  Carrock Fell, 141, 146
  Cassiar asbestos, 102
  Closehouse, 148
  Dolcoath, 150
  Force Crag, 158
  Foxdale, Isle of Man, 144
  Geevor, 155
  Greenside, 141, 158
  Gortdrum, 92
  Kilembe, 83–107
  N'changa, 104
  Parys Mountain, 141, 146–7
  Roan Antelope, 99
  Settlingstones, 145
  South Crofty, 155
  Strontian, 149
  Thetford, 93
  Thornthwaite, 141
  Witwatersrand, 99, 303
mine design, 92, 96–7
mine workings, old, 67, 68, 271, 322
mineral processing, 99
  rights, 24
  zonation, 157
minerals industry, 80–3
mineralogy, 91–2, 101
mineralizing fluids, 154, 160–3
mining, 298, 302
Mississippi Valley deposits, 147
Mohole, 9
molybolenite, 140
Mont Toc, 286
montmorillonite, 171, 206
motorways, 328–9
moulding sand, 173
mudflow, 303

National Coal Board (NCB), 11, 31, 54, 55, 62, 65
National Parks, 221
National Nature Reserves, 330–3
natural gas, composition, 40
Nature Conservancy Council, 313, 330
  Reserves, 330–3
Netherlands, 252
New Mexico, 243
nickel, 79, 146
Nile Valley, 256
Norfolk Broads, 310
North Sea, 323
Northfleet, 208
Nubian Series, 256–7
nuclear power stations, 273–5

nuclear resource evaluation, 20

oases, 256
offshore aggregates, 173
Ogof Fyfnonn Ddu Reservoir, 332
oil, 1, 2, 8, 9, 33–53, 296, 299
  migration, 45
  origin, 38
  production, 22–4, 52
  recovery, 15, 36–7
  reservoirs, 34–6, 41
  reserves, 10, 15, 18, 42–3
  resources, 50–1
  sands, 10, 20–1
  shales, 18–19, 27
  traps, 45–6
oil and gas fields
  Abu Dhabi, 38
  Alaska, 26
  Algeria, 31
  Anadarko, 8
  Arctic Canada, 10
  Athabasca oil sands, 10, 20, 27
  Bahrain, 8
  Britain, Midlands, 51
  Gulf Coast, 8
  Holland, 51
  Iran, 27
  Lake Maracaibo, 9
  Middle East, 15
  Mississippi delta, 8–9
  North Sea, 9, 22, 31, 47, 51–2
  Norway, 52
  Orinoco Tarbelt, Venezuela, 10
  Persian Gulf, 8
  USSR, 15, 44
opaline silica, 170, 172
opencast pits, 65, 99, 104, 298
Opencast Executive, 66–9
Ordinary Portland Cement, 196
OPEC, 31, 33
Orange Free State, South Africa, 212
ore, 101, 137
  deposits, 1, 83, 142–3, 152–3, 154
  minerals, 5
  reserves, 90–1, 97
orebodies, 85–8, 103
Oxford Clay, 323
oxidation, coal, 74

Pakistan, 255
Palaeozoic, 44
paraffins, 37
peat, 310
pegmatites, 85, 88
periclase, 215
permeability, 34, 41, 45, 229–30, 232, 239, 297, 302
permafrost, 26, 28
petroleum engineering, 49

petroleum storage, 250
phosphorus, 217
piezometric surface, 226
piles, 273–5
pipelines, 28, 210, 328
pits, after-use, 322–5
planning, 105
plate tectonics, 9, 125, 146, 290
Pleistocene, 224, 252, 292
PLUTO, 28
pozzolans, 197
polished stone value, 179–80
pollution, 26–8, 95, 221, 224, 248–50, 251,
   289, 321
Pompeii, 288
pore pressures, 295–6, 300–1
porosity, 34–6, 41, 193, 228–9, 239, 290–2
porphyry coppers, 78, 81, 91
Portland Cement, 190–7
potash, 96
Pre-Cambrian, 85, 125, 140, 198, 301, 303
precipitation, 233–4
project reconnaissance, 125–9
pulverized fly-ash, 180, 185, 324
pumping test, 239–40
pyrite, 57, 85, 99, 170, 172, 193
pyrrhotite, 90, 99, 140

qanats, 237
quarries, 315, 319
   old, 221
   waste, 189
Quaternary, 296
Quattara Depression, 256
quick clays, 292

raft, 274
Recent, 252
recharge, 230, 233, 235, 240, 255–6
refractory bricks, 191
remote sensing, 7–8
reservoirs, 273, 297–8, 329–30
resistivity, 264
resource evaluation, 4–5, 16–17, 66, 70–71, 98,
   102
Rhaetic Bone Bed, 328
Rhine Delta, 252
rip-rap, 181
road aggregate, 177–8
roads, 176–7, 282–3
rock mechanics, 99, 104
   bursts, 299
   slopes, 80, 83, 97
   strength, 192, 278
   quality designation (RQD), 277, 268–9
Rockall, 25
rockfill, 182–3
Rodovia dos Imigrantes motorway, 282–3
rubble, 181
Ruhr Valley, 254

runoff, 233–4, 304

saline intrusion, 237, 240, 243, 251–2
   groundwater, 160–61, 226, 247–8
salt caverns, 29
   domes, 8, 296
sampling, 91, 102, 214, 265–6
sand, 94, 186
sand and gravel, 166, 175, 168, 320
sandstone, 34, 41
Scandinavia, 292
seabed geology, 28
seatearth, 60
sedimentary rocks, 79, 180
seismic disturbances, 301
   investigation, 264
semi-dry process for cement, 201, 206, 210
shales, 207, 278
shear zones, 88, 94, 103–4, 301
silica ratio, 203, 205–7, 217
   sand, 212–13
sinkholes, 303
skid resistance, 179–80
slags, 101, 180, 184–6
slate waste, 188
slope instability, 71–2, 97
Smith, William, 54, 327
Snowy Mountains, 29
soil moisture, 293
South Africa, 29, 129, 131
specific retention, 228
   yield, 228
specification, 271
spread footings, 321
stable isotopes, 162
staging of site investigations, 261–2
stoping, 103
storage, 240, 296
   coefficient, 228, 241
stratigraphy, 41
structural control of oil deposits, 44, 46–7
   of ore deposits, 88–9, 93, 149–50
subsidence, 99, 104, 243–4, 289, 296–7, 299,
   300–1, 303
Sudan, 256
Suess, Edward, 140
sulphate, 246, 248
sulphides, 78, 113, 120, 140, 146, 246
super-quarries, 194
Swansea Valley, 310

tar sands, 18
Tasker Quarry, Shropshire, 319
taxation, 84
team, work of geologists in, 3–4, 98, 126, 259
temperature control of ore deposits, 156–60
tensile strain, 302
Tertiary, 146, 241, 243
testing, 270
Texas, 243

Index

Thames, 207–9, 252
thermal waters, 2
tin, 78, 137, 140
tips, spoil, 318–19, 321–4
Tokyo, 296
tonstein, 57
townships, 93
toxic chemicals, 314
trace elements, 95
transmissivity, 229, 239, 251–2, 256, 297
travertine, 193
trend surface analysis, 220
tricalcium aluminate, 198
    silicate, 198, 205
tropical weathering, 94
tundra, 26–7
tungsten, 141
tunnels, 225, 237, 277–80, 295, 327
Tyne Vehicular Tunnel, 284

Uganda, 84, 212
underground storage, 29, 49
    stress conditions, 59

United Nations, 84
United States of America, 212
uranium, 9–12, 84, 148

Vancouver Island, 222
veins, 141–4, 148, 151, 154
volcanic deposits, 212

wallrocks, 148–9, 155
water, 33, 46, 59, 95, 222, 314, 322
    drive, 15
    flooding, 300
    table, 35, 213
waste, 187–8, 212, 248, 302
    disposal, 184–5, 221, 226
weathering, 131, 225, 282–4
well disposal, 301
wet process for cement, 201, 206, 210
Wren's Nest, 331

Zambia, 82
Zechstein Sea, 146
zonation, mineral, 158–9